Δ $^{20}/_{2d}$ Sci   £2.50

D1630583

The
Barmaid's Brain
and other strange tales
from science

The
Barmaid's Brain
and other strange tales
from science

JAY INGRAM

AURUM PRESS

First published in Great Britain
1999 by Aurum Press Ltd
25 Bedford Avenue, London WC1B 3AT

This paperback edition first published in 2001 by Aurum Press

Copyright © Jay Ingram, 1998, 1999

ISBN 1 85410 755 0

First published in Canada by Viking, a division of the Penguin Group

All rights reserved. No part of this book may be reproduced or utilized in any
form or by any means, electronic or mechanical, including photocopying, recording or by any information storage and retrieval system, without prior permission
in writing from Aurum Press Ltd.

A catalogue record for this book is available from the British Library.

Copyright Acknowledgements
Illustrations from the following were used by permission: Figure on page 239
from Volume 7, page 113 of the *Royal Institution Library of Science*. Courtesy of
the Royal Institution. The walrus image drawing on page 25 is from *NATURE*
(January 29, 1981). Courtesy of *NATURE*. The illustration of bacteriophage on
page 203 is used by permission of Frederick A. Eiserling. The Vinland map on
page 152 is used by permission of the Beineke Rare Book and Manuscript
Library, Yale University. All attempts were made to secure permission for other
illustrations.

10 9 8 7 6 5 4 3 2 1
2005 2004 2003 2002 2001

Printed in Great Britain by MPG Books Ltd, Bodmin

To Cynthia, Rachel, Amelia and Max

Contents

Acknowledgements ix
Introduction xi

HUMAN BEHAVIOUR

I Just Had to Laugh   3
Seeing Things   18
Sane in an Insane World   27
The Barmaid's Brain   40

CURIOSITIES OF LIFE

The Invention of Thievery   55
The Plant That Rolls   67
Consumed by Learning   77
Why Do Moths Fly to Lights?   87
Homo Aquaticus   102

SCIENCE AND HISTORY

Saint Joan   121
The Effect of Witchcraft on the Brain   138
The Vinland Map   150
The Burning Mirrors of Syracuse   165
The Monks Who Saw the Moon Split Open   174

NATURAL BATTLES

Antlion King   187
The Bacteria Eaters   199
An Uneasy Bargain   210
A Silver Lining   222

HOW THINGS WORK

Tee Time at the Royal Institution   233
It'll Practically Go Forever   242
Going Up?   254

Bibliography 265

Acknowledgements

There were several groups of people whose help allowed me to write this book. The first and most important is the small group of Cynthia, Rachel, Amelia and Max. They are fun to be around and a constant reminder of what's important.

If I hadn't been able to turn to excellent researchers, I might never have finished this book. Most of it was more than capably handled by Sue Gemmell and Laura Boast. Whenever either took off for the library, or started e-mailing, you knew something good would result. A writer can't ask for anything more. Louise MacLeod, who works with me at Discovery, was their illustrious predecessor. We can't remember exactly what she did for this book but she deserves a mention nonetheless. There are others at Discovery, like Penny Park, Nancy Block, and Margaret Polanyi, who helped too. And who could forget the illustrious Paul Lewis who gave me time off at a critical point in the writing of this book.

Of course, Meg Masters and Mary Adachi, as usual, turned a first-draft manuscript into a book, an achievement on about the same level as humankind progressing from hand signals to articulate speech. I must say though that Mary provided better food than Meg did.

Finally there were a number of scholars and scientists who provided me with important pieces of information or the leads to them. Prominent among them were Martin Moskovits, Bert Hall and Dennis Proffitt, but I am also indebted to Suzanne MacDonald, Ron Thomson, David Kirk, David Sherry, Louis Lefebvre, Rhanor Gillette, Alan Kingstone, Conrad Berube, John Langdon and Betty Kutter.

Introduction

This book is a selection of my favourite science stories. Some I have been keeping in my files for years, waiting for an opportunity to do something with them. Others are new. It is a diverse collection, but there are some common threads among them.

One is that most of these stories are far from the mainstream of science; the scientists involved in them are not likely to be travelling to Stockholm to receive a Nobel Prize. This is not to say these are accounts of bad science—they aren't. Nor are they fringe science of the UFO/sasquatch type, where too often paranoia and ideology override good sense. Most of these stories are drawn from the *edges* of science, unfamiliar territory for researchers, where practising science poses unusual challenges. Out at the edges researchers don't isolate genes; they try to understand why those genes exist. They try to read the minds and actions of those long dead. They sometimes try to answer the simplest questions to which mainstream science has paid little attention, like why moths are attracted to lights.

The scientists who choose to do this kind of research often reveal more of the subjective side of science than we usually see. Leaps of imagination, strong biases, intolerance, sharp criticism— all are part of science, but in these stories they are painted in bold colours.

A second theme common to all these stories is that the final chapter is never written. Trying to explain the strange and wonderful in this world is an endless process: each new, closer look brings surprises. That is what makes these stories entertaining; that's why these are some of my favourites.

The Barmaid's Brain

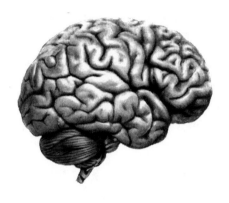

Human Behaviour

I Just Had to Laugh

There are few things as funny as reading psychologists' or doctors' descriptions of laughter:

The primary component of laughter is an abrupt, strong expiration at the beginning, followed by a series of expiration-inspiration microcycles with interval pauses.

The true manifestation of laughter is . . . the abrupt expiration due to a sudden contraction of the intercostal muscles.

A laugh is characterised by a series of short vowel-like notes (syllables), each about 75 milliseconds long . . .

It's enough to set you off on a series of expiration-inspiration microcycles with interval pauses. But these techno-speak descriptions, as remote from reality as they seem, are the first step in defining, then trying to understand what laughter is and why we do it. The problem is that because we are so familiar with it (it has been claimed that there is no record of anyone who has not laughed at least once in his/her life), we overlook the fact that it is a peculiar habit and a fundamentally mysterious part of our behaviour. As psychologist Robert Provine (who's one of the few academics to study laughter) wrote: "What would [an alien] make of the large bipedal animals emitting paroxysms of sound from a toothy vent in their faces?"

When Provine performed some detailed acoustic studies of laughter, he uncovered some unexpected links to speech. He took recorded samples of laughter to the National Zoo in Washington, DC, where they were analyzed with a sound spectrograph, which produces images of the frequency and intensity of animal vocalizations. The resulting data prompted Provine to describe laughter as a series of vowel-like sounds repeated at regular intervals: ha-ha-ha-ha. These are all about the same length. But the vowels themselves can change: ho-ho-ho-ho. Usually a person laughs roughly the same way all the time—I bet you could identify the laughter of your friends sight unseen, even if the pitch were altered. Provine pointed out something that might seem obvious but nonetheless needs to be said: two different laugh-vowel sounds are unlikely to appear together in the same laugh. Santa never laughs ho-ha-ho-ha. If there are intrusions of a different sound, they occur only at the beginning or the end of the laugh.

That's not all. These experiments also revealed that laugh notes, like the notes of the singing or speaking voice, consist of a fundamental frequency (or pitch) accompanied by a number of harmonics or multiples of that frequency. The fundamental frequency of male laughter is almost an octave below that of females; the female frequency is close to the C above middle C on the piano, while the male frequency is roughly the D next to middle C.

Robert Provine argues that while there is tremendous variation within laughter, all human laughs follow this pattern.

Provine found that the spaces between the laugh vowels contribute very little to the sound: if they're removed, leaving only "ha's" and "ho's," the laugh still sounds almost exactly the same. The punctuating silences themselves, when strung together, amount to nothing more than a prolonged sigh. One final point from these sonic investigations: while the individual notes of laughter sound much the same played backwards or forwards, an entire sequence of backward laughter sounds weird because it rises in volume, instead of fading out as it does normally as we run out of breath.

Even given the relatively constant timing and sound qualities of laughter, there is obviously still room for tremendous variation. In response to a bad joke, you barely make it over the laughter threshold (although the social situation might demand that you laugh as best you can), but you can also "dissolve" into laughter so powerful it leaves you bent over and weeping.

The anatomy of laughter represents only one small part of the research on the subject. Social psychologists and psychiatrists have emptied a few inkwells with their speculations on the functions of laughter. Freud—naturally—weighed in with a theory that laughter represented a discharge of built-up sexual tension, and while there aren't many who agree with him about the sexual part, there is general agreement that laughter results from an abrupt change in the direction of thought or a sudden reduction in tension or both. A joke leads you along one path, often gravely serious, then suddenly reverses with the punchline. Slapstick juxtaposes a man in a suit with a cream pie in his face. The classic example of tension reduction is the movie portrayal of the mob boss and his boys sitting around the table. He smiles, they do the same. He laughs, they laugh. He stops, they stop. He laughs heartily, they fall all over themselves in hysterics.

There are entire books written about what triggers laughter and why; there is nothing dramatically new on that front. However, there has been research in recent years on the brain mechanisms underlying laughter and even some guesses as to how this unique behaviour came to be.

Searching the brain for a "laugh centre" isn't easy. Everybody laughs, so you can't compare laughers and non-laughers; brain imaging requires the person being imaged to stay relatively still, a great challenge when you're laughing. But in any kind of brain research there are individuals with some kind of brain damage that specifically affects the behaviour you're looking for; in this case, the damage causes what's called "pathological laughter."

One of the most recent cases involved a twenty-three-year-old male ensign in the US Navy who was at the controls of a jet trainer when his instructor heard what he described as "uproarious" laughter. The aircraft began drifting towards another flying in the same formation and the instructor seized the controls to prevent a mid-air collision. The pilot was assumed to have a peculiar form of epilepsy and was sent to a neurologist.

The ensign explained that the symptoms had begun about eighteen months earlier: his family and friends started telling him that he was laughing in his sleep, sometimes so loudly he would wake himself up. What was worse, at least socially, was that he would begin laughing loudly at inappropriate times during the day, like at officers' meetings. Apparently these spells of laughter could happen at any time, and lasted about ten seconds. They were accompanied by a vague feeling of lost concentration; sometimes this feeling occurred without the laughter. After about ten seconds, he would snap out of the seizure and go on as before. Most significantly, the patient reported that his laughter was not accompanied by any feelings of mirth; in other words, the act of laughing was some sort of automatic act—a seizure—triggered independently of his thoughts and actions at the time.

The patient was put on an anti-seizure medication and apparently has never suffered a laugh attack since, although for some time he continued to experience the strange aura that accompanied the laughter. This case suggests strongly that there is some sort of laughter centre in the brain, the activity of which was being triggered in these seizures. Although the patient once had a seizure while doctors were giving him an EEG, an electroencephalogram, it wasn't possible to pinpoint the location of any specific site that might have been involved.

Laughter seizures like this have been reported before but are rare: they represent less than 1 per cent of all cases of epilepsy, and when taken together provide only a vague indication of which places in the brain might be involved in the generation or control of laughter. Most patients with these unusual seizures, like the laughing pilot, experience no joy or merriment with their laughter.

In another case, reported in 1990, an employee of a fast-food restaurant arrived at the emergency ward in Akron, Ohio, laughing uncontrollably. Apparently he had been spraying an insecticide when the wind changed and he inhaled a full breath of the compound WC Insect Finish. Within seconds he had started to laugh and couldn't stop. He felt some numbness and tremor but was otherwise completely normal. After an hour and forty minutes of laughter he began to complain of pain in his abdomen. When he was given a shot of Valium, the pain, tremor and laughter stopped, never to reappear.

The most common causes of pathological laughter are brain diseases that eventually disrupt centres that control a variety of normal movements. So in some cases patients begin to lose control of their chewing, speaking, sometimes even breathing, but at the same time are capable of crying, swallowing or laughing. Such patients may laugh either disproportionately loudly or at inappropriate times. Even the laughter itself can assume a disturbing quality: "prolonged and distorted like a scream or a wail."

There isn't much that can be said from these disparate cases other than there are likely multiple neural pathways governing laughter. Some must be inhibitory, and when those are damaged patients laugh loudly and inappropriately. Some are excitatory. When those pathways are triggered abnormally, as by an insecticide, laughing can't be turned off. But all of this is pretty vague— practically any human behaviour involves inhibitory and excitatory pathways in the brain.

However, early in 1998 a case of laughter was reported that shed new light on what goes on in the laughing brain. A group of researchers at UCLA led by Dr. Itzhak Fried were exploring the brain of a sixteen-year-old girl who was suffering from intractable epileptic seizures. Their intent was to find the scarred or abnormal part of the brain that was triggering the girl's seizures, then remove it surgically. To do that they had exposed the surface of her brain and were stimulating a variety of places with an electrode—this kind of surface mapping has been standard pre-surgical practice since Wilder Penfield did it in the 1940s and 1950s. Just as Penfield might have reawakened long-lost memories with the tip of his electrode (although that claim is controversial), the UCLA team came up with their own surprise. Whenever they stimulated a part of the brain called the supplementary motor area, at the front of the brain, the girl laughed.

This was a curious finding for more than one reason. The supplementary motor area is a part of the brain that is a sophisticated neural add-on (at least in evolutionary terms) to the so-called primary motor cortex, a strip of tissue at the top of the brain that issues commands for movements. The supplementary motor area refines those movements by organizing, planning and helping to execute complicated sequences of them, such as playing the melody of a Bach fugue on the piano or uttering a complicated sentence. The laughter-sensitive area was close to the area responsible for speech, but it seems odd that this part of the brain should be recruited for the genesis of laughter when you consider that the sounds of laughter are not complicated, but in fact are relatively stereotyped and repetitive. Nonetheless, touching the electrode to this two-centimetre-square piece of brain invariably caused the girl to laugh.

But this finding wasn't as intriguing as the girl's reaction to her own laughter. Unlike patients who laugh because of brain damage, the girl at UCLA seemed to have the full laughter experience: she experienced mirth, and even more than that, claimed to be able to identify the "funny" event that caused her to laugh. So for instance, when she was looking at pictures to test her ability to name objects, and the electrode touched the laughter centre, she exclaimed, "the horse is funny"; once she just looked at the doctors and said, "you guys are just so funny…standing around." The only reasonable explanation for this strange labelling of unfunny events as laughter-provoking is that the girl's brain was making it all up, confabulating to explain (to itself) why there was laughter.

This isn't such an outrageous idea. Several years ago neuroscientist Michael Gazzaniga proposed that the left hemisphere of the brain contained something he called the "interpreter," a neural centre whose job it is to make sense of activity in other parts of the brain. Gazzaniga was attempting to explain some of the weird moments he had witnessed when working with split-brain patients, people who have had the connection between their two hemispheres severed in order to limit the spread of seizures. Gazzaniga had often seen the left hemisphere fabricate explanations. For instance, the command, "walk," was given to a patient's right hemisphere—and that hemisphere only. When, moments later, the patient got up to "walk," the question, "What are you doing?" was directed to his left hemisphere. The left hemisphere, able to respond verbally because it contains the speech centres, produced the reply: "I'm thirsty, I just thought I'd go down the hall and get a Coke." First came the behaviour, then the rationale.

So when the sixteen-year-old laughed, then remarked on how funny the assembled doctors were, her brain was apparently doing the same thing. The UCLA group suggests that a joke, the resulting feeling of mirth and the act of laughter are all stations along a neural network in the brain. While normally the joke comes first, this case illustrates that it is possible to produce the laughter first, which then triggers the rest of the network. However, you have to be cautious here: when you consider the number of different situations which can provoke laughter, from tickling to jokes to nervousness, and the complexity of the muscle movements involved, there is no way that the brain pathways governing laughter can be anything less than extremely complicated.

The implications of the UCLA experiment are profound—especially for stand-up comics. If a comic can incite the people in the audience to laugh—for whatever reason—their brains will likely attribute that laughter to the comic's jokes. Canned laughter in television sitcoms has been used for nearly fifty years now and researchers have confirmed that it both makes viewers laugh more and persuades them that the show is funnier, two quite different conclusions. Robert Provine has suggested that our brains might contain laughter detectors which are linked more or less directly to laughter generators, so that hearing laughter would make us laugh, even in the absence of anything funny. But such a neural automatism wouldn't by itself convince viewers that laugh-tracked programmes were actually funnier—that would only happen if there were additional brain mechanisms whose job it was to identify causes for that laughter. It appears from the UCLA experiment that such brain software does exist.

Laughter is contagious, but only to a point. I have seen claims that laughter by itself—accompanied by nothing that could be construed as a joke—does not provoke laughter in others, but there are recordings of both laughter alone, and laughter with music, that seem to have the effect of making the listeners laugh. On the other hand, Robert Provine found that if he played an eighteen-second recording of laughter over and over again for students, they laughed less and less in response, until by the tenth time almost no one laughed and most of them described the laughter as obnoxious. I'm sure the social psychology of laughter will turn out to be as complex as its brain mechanisms.

There is one example in the medical literature from 1963 of the contagiousness of laughter run amok. The report is titled *An Epidemic of Laughing in The Bukoba District of Tanganyika.* Hundreds of people, mostly adolescent girls (but also boys and adults of both sexes), succumbed to an apparent epidemic of laughter. It began in January 1962 in Kashasha village near Lake Victoria (in what is now Tanzania). Children would suddenly start laughing and crying, sometimes for hours at a time. There were no physical symptoms: no fever, no tremors, no fainting and certainly no fatalities. The laughing attacks spread from one afflicted person to the next—at the mission school in Kashasha, 95 out of 159 pupils were affected over a period of about seven weeks. The school closed in March, and a few days later, the epidemic of laughter broke out in some of the villages to which the girls from the school had been transferred. It was clear that the laughter spread from person to person, but no evidence of any infectious agent, toxic substance in food or physical abnormality in the patients was ever found. All laboratory searches for abnormalities in the blood, viruses or bacteria came up empty, and all the investigating doctors could conclude was that it was a case of mass hysteria. It still ranks as one of the most peculiar cases of pathological laughter and the most extreme demonstration of the contagiousness of laughter.

None of what we know about laughter, the multiple sites in the brain controlling it, its contagiousness and its peculiar repetitive form tell us anything about why we do it. In that sense, although laughing is intimately bound up with conversation, it is more like crying (Why shed drops of water from your eyes to express emotion?) than speech.

There have been some attempts to explain why we laugh, as opposed to why we find something amusing. A better way to put it might be: Why do we respond to humour by emitting bursts of weird sounds? Nobody really knows of course, but there are some interesting speculations on the matter.

One that seems to come right out of left field has been proposed by Vilyanur Ramachandran of the University of California at San Diego. Left field in this instance is the unusual brain-damage syndrome called anosognosia, the lack of awareness or even denial of a body part. This is sometimes seen in patients who have recently suffered a stroke that has paralyzed one side of the body. I related the details of one such case in *The Burning House:* a woman had suffered a stroke that paralyzed her left side, and while in hospital came to believe that the left arm that was clearly attached to her body was not actually hers. She believed— honestly and completely—that it had been the left arm of the person who had occupied her bed before her. She was able to back up this belief by pointing out details that proved to her it wasn't her arm, such as the hospital bracelet with a name different from hers by one letter (apparently she was misreading it), thicker fingers on her hand and a wedding band that seemed like hers but different in detail.

Vilyanur Ramachandran begins his theory of laughter by

detailing several cases like this, some of which involved simple denial that paralysis existed:

"Can you point to my nose with your right hand?"
(The patient points)
"Mrs. D, point to me with your left hand."
(Her hand lies paralyzed in front of her.)
"Mrs. D are you pointing to my nose?"
"Yes."
"Can you clearly see it pointing?"
"Yes, it is about two inches from your nose."

It might seem obvious that such patients are simply protecting themselves from the awful reality of what has happened (and in the case I cited the patient eventually came to understand that it was indeed her arm), but Ramachandran points out that it never seems to happen when the damage occurs to the left hemisphere of the brain (thus paralyzing the right side of the body). Anosognosia appears when the right hemisphere is damaged and the left is intact. And it is that hemispheric difference that Ramachandran thinks is the key to both this syndrome and laughter.

He argues that the left hemisphere has the responsibility for maintaining a thread of consistency in our behaviour, connecting moment-by-moment experiences in a sensible way. This is an important role because among the thousands of sensory impressions gaining access to our brains are some that don't fit with your concept of the world and your place in it. It's the job of your left hemisphere to screen those out and maintain what Ramachandran calls "the script." The right hemisphere, on the other hand, has the opposite role: when some important new piece of information comes along, it is the job of the right hemisphere to persuade the left that this information, rather than being disregarded, should be incorporated into the brain's script. Ramachandran thinks of the right hemisphere as the devil's advocate. The role he assigns to the left hemisphere is reminiscent of Michael Gazzaniga's left-brain interpreter.

In anosognosia, the right hemisphere—the devil's advocate—has been disabled, and the left hemisphere is now free to go to the most absurd lengths to cling to the script. It will ignore even the most dramatic anomalies. It will not even acknowledge paralysis, to the extreme degree of asserting that a paralyzed arm is pointing to a nose, or that it doesn't belong to the patient at all.

Ramachandran claims that humour and laughter can also be explained by these hemispheric differences, based on the well-established notion that a joke leads you along one path, then suddenly surprises you with a trick ending. In this case both hemispheres are functioning normally: the left follows the story as it builds up, while the right fulfils its usual role of looking for anomalies. The punchline is exactly the sort of incongruity or unexpected twist that the right hemisphere seizes on, but in the case of a joke, it turns out to be trivial. So the tension that has built up (and always does when something truly novel and perhaps threatening happens in real life) must be dissipated, and we accomplish that by laughing. Ramachandran even goes on to suggest that the loud repetitive nature of laughter can be traced to our evolutionary past, when it was an alert to others that something strange—which every right hemisphere in the group had been tracking—had turned out to be trivial. I laugh and the rest of you can relax: that rustling wasn't a rattlesnake, it was a dung beetle.

(Our closest relatives, the chimpanzees, laugh as well, but do it much differently than we do, by inhaling and exhaling rapidly. Although this is not at all the way we laugh, it might still be explained by Ramachandran's theory.)

I am intrigued by these ideas, and it is easy to see how it explains the laughter of relief when something you fear turns out to be harmless. But does it explain why an infant laughs (or at least gurgles) when an adult says, "Goo goo ga ga"? Nor is it clear how Ramachandran would explain what the late Benjamin Spock described as the first joke ever told: the passing of wind. As any parent knows, that general subject provokes uproarious laughter at any time in any situation. Maybe there are some specialized brain circuits involved.

I can't leave the subject of laughing (and why we do it) without addressing tickling (and why we can't do it to ourselves). And wouldn't you know it—when something is biological you can almost guarantee that Charles Darwin had something to say about it, and that what he said still has a grain of truth in it. In 1872 Darwin drew an analogy between laughter and tickling: "The imagination is sometimes said to be tickled by a ludicrous idea; and this so-called tickling of the mind is curiously analogous with the body . . . The touch must be light, and an idea or event, to be ludicrous, must not be of grave import."

Darwin observed that people who are ticklish are also people who laugh easily for other reasons. They tend to get goosebumps, to blush and to cry. He suspected these physical phenomena were somehow related. In 1990 Alan Fridlund and Jennifer Loftis, psychologists at the University of California at Santa Barbara, tested these ideas and found them to be true: blushing, laughing, ticklishness, crying and goosebumps go together. Fridlund and Loftis point out that laughing and crying are sometimes inter-changeable ("I laughed until the tears rolled down my face") and that they might both be ways of releasing the tension that is obvi-ously associated with blushing and goosebumps. And although they think there must be a strong genetic component to these related behaviours, they also point out that the environment plays a role. The parental habit of tickling infants to make them laugh establishes the connection between them (and sometimes crying) early in life. The authors did admit that much more study would have to be done.

But while tickling appears to be part of an emotional package, it is curious in and of itself: we want it to end as quickly as possible but we're laughing ourselves silly at the same time. It is probably that need to escape that explains why we can't tickle ourselves: when someone else is doing it we can't be sure they will stop. However, a shorthand explanation like that isn't good enough for psychologists who have actually tried to answer the question, "Why can't we tickle ourselves?" in the lab.

Nearly thirty years ago a team of British psychologists published a report in the journal *Nature* called "Preliminary Observations on Tickling Oneself." Rather than speculate, they built a tickling box, a complicated thing about the size of a shoebox with a slot in the top. A plastic pointer could be moved along the slot by a handle, and it would tickle the sole of a foot placed on the box. The pointer was cleverly counterweighted so as to exert the same pressure no matter how it was being moved.

Thirty undergraduate students were tickled in different ways: either the experimenter controlled the pointer, or the student did; or the student held the handle passively while the experimenter moved it. The results were more or less what you'd expect. Self-tickling didn't work; being tickled by the experimenter did, but holding the handle passively had an effect that was in between.

A few years later Guy Claxton at the University of London confirmed these findings by showing that subjects were more ticklish if they were being tickled by someone else (five light strokes with a feather in five seconds) than if they held the feather in their own hand but had the hand moved by another person. Apparently the sensation of movement reduces the ticklishness even if the subject isn't in control of the movement. He also found that subjects were more ticklish if they had their eyes closed—the surprise tickle—something that Darwin had suggested. As I was writing this chapter, a new experiment was reported that produced images of the brain during tickling. Those images revealed that, in the case of self-tickling, the cerebellum, a part of the brain responsible for co-ordinating movements, is activated. Furthermore, the cerebellum might then warn other parts of the brain that a tickle is on its way, dramatically reducing the sensation. When another person does the tickling, the brain's cerebellum is inactive and no warning is sent.

However, the children I know laugh harder when they know exactly where and when they are going to be tickled—the tickle of anticipation. Maybe in their case the cerebellum's warning heightens, rather than dampens, the tickle. Why children are different is not yet well understood.

That last phrase is a good description of the scientific understanding of laughter. There are suggestions, hints, some experiments that shed light on the complex relationship between humour, tickling, laughter and the brain, but we are still far short of understanding why we respond to funny situations (and sometimes ones that aren't funny at all) with repetitive blasts of air. It almost brings on abrupt expirations due to a sudden contraction of the intercostal muscles.

Seeing Things

————————
————————

It is startling to realize just how much we rely on one sense out of five, our vision. That reliance is so all-pervasive that it has infiltrated our language: "I don't see what you're getting at" . . . "Do you get the picture?" . . . "Those stories are ruining her image." We may consider ourselves deep thinkers, but neuroscientists have established that more than half the human brain is specialized simply for processing visual information. It is a huge amount, but the visual parts of our brains have an enormous task to perform: as you cast your eyes around you, every split-second scene is broken down to different features like colour, movement and form. Each of those is analyzed independently. Finally those features are recombined and we (somehow) become aware of what we are looking at. Then a new scene appears.

Those areas of the brain concerned with vision have the challenging task of making sense of the deluge of photons of light impacting on the two retinas every moment, a process that starts with visual pigments in rod and cone cells being altered physically by the energy of a photon and ends with your recognition of your grandmother's face. It's no surprise that the brain takes shortcuts and makes assumptions. Ninety-nine per cent of the time they're worth taking; one per cent of the time those shortcuts take the brain down the wrong road. When that happens you experience an illusion.

For instance, it's sensible for the brain to assume that the larger the image, the closer the object. From that moment hundreds of millions of years ago when being eaten by some other creature became a possibility, eyes and brains have found that to be a life-saving shortcut. Animals that reacted quickly to a looming image left far more offspring than prey whose visual systems spent a few extra moments trying to establish a positive identification. But it is not always true that image size correlates with distance.

It is easy to draw a scene in which the images of distant people are the same size as those of people in the foreground. The human brain never sees anything like that in real life, and it is fooled into thinking that the distant people are huge. Such tricks are commonly known as "optical" illusions, but because they are tricks of the brain rather than the eye, they are more accurately called visual illusions.

The better understood the process of vision is, the easier it is for psychologists to invent illusions that take advantage of the visual brain's hard-wired assumptions. However, it is not just in the lab that the visual brain can be fooled. One of its primary assumptions is that light travels in straight lines. For all intents and purposes, over distances that concern most living things—a few metres or less—light does travel in straight lines. But we can see much farther than that and when the objects we're looking at are as much as a few kilometres or so away, strange things can happen to the light en route from those objects to our eyes. The result is a mirage.

The Barmaid's Brain

The archetype is the desert mirage, the appearance of what looks like sheets of water covering the sand in the distance. You encounter the same effect when you see what looks like water on the road in the summer. More on what causes them in a moment, but there are much more spectacular mirages than these two.

For instance, the people of Buffalo, New York, were shocked on August 16, 1894 when from ten to eleven o'clock in the morning a detailed image of downtown Toronto appeared in the sky. *Scientific American* magazine reported that Toronto church spires could be counted, despite being more than fifty miles away. The people of Ramsgate, England, were surprised once when Dover Castle appeared to rise above the hill that usually hides it. The most amazing of such mirages was reported by Captain John Bartlett from his ship the *Effie M. Morrissey* in the North Atlantic on July 17, 1939. Captain Bartlett identified the outlines of a 1430-metre (roughly 4700-foot) peak on the west coast of Iceland, more than 500 kilometres away.

None of these people were imagining what they saw: these images were created because the light bent as it travelled though the air. Light would travel in straight lines if the temperature everywhere in the air was the same, but if for any reason layers of air of different temperatures come together, whether it's cold air trapping warm beneath it or warm air lying above cold, mirages can result. In cases like this, the light travelling through these layers of warm and cold will bend, always curving towards the colder side.

So if there is a layer of warm air immediately above cold air near the ground (as can happen over a lake in early summer), light moving towards your eye from a distant object will bend down, with the result that objects that would normally be over the horizon now come into view. If the change in temperature with altitude is great enough, light will bend to the same degree as the earth curves, creating what looks like a perfectly flat earth with a horizon extending out to infinity. Images of distant objects are brought to your eyes and you become Captain Bartlett, seeing modest peaks 500 kilometres distant. If the light bends even more, the horizon can seem to lift up as if you were standing in the middle of a shallow saucer. Explorer David Thompson wrote of the experience while on his travels through western Canada: "On one occasion, going to an Isle where I had two traps for Foxes, when about one mile distant, the ice between me and the Isle appeared of a concave form, which, if I entered, I should slide into its hollow, sensible of the illusion, it had the power to perplex me. I found my snowshoes, on a level, and advanced slowly, as afraid to slide into it . . ."

In the reverse example, when cold air overlays a blanket of hot air close to the ground, light rays will bend upward (always towards the lower temperature) and the horizon seems to have approached you, leaving you standing, not in a saucer, but on the surface of a shrunken and more highly curved Earth.

There are more elaborate mirages, like the famous *fata morgana*, where the light rays are so disturbed as they travel through a turbulent atmosphere that images are stretched, multiplied and blurred, creating what literally appear to be castles in the air. Even the desert oasis/wet highway mirage is slightly complicated: in this case the hottest air is next to the road (or the sand in the case of the desert). In certain circumstances the light rays bend to the point of looping, and images can appear twice. In the highway version the sky appears once above the road, and once, inverted, below the horizon on the road. The apparent distance to the mirage depends on the degree of bending and the height of the observer. This of course means that if you actually are crawling across the desert on your hands and knees looking for water, you will see a better mirage—the most dramatic visual effects are seen from ground level. But no matter how vivid the mirage, it is true that as you move forward, it recedes.

Just experiencing a mirage, especially one of the truly dramatic ones is fantastic enough, but a man named Waldemar Lehn at the University of Manitoba thinks that for the ancient Norse, mirages were much more than casual entertainment. He contends they may have played a critical role in the Norse exploration of the North Atlantic and in the tales that were told about those voyages. The story of the discovery of Greenland by Erik the Red in AD 981 is a perfect example.

Erik's life was a tumultuous one. Even in a society where lives were taken right and left, Erik stood out. After being involved in several murders in Iceland, he was banned by the legislative assembly for three years. He had nowhere to go but to sea, and although the accounts differ in detail, he apparently vowed to sail to the west, either to rediscover a land that had been visited once fifty years earlier, or to investigate claims that there was something far off to the west of Iceland that was neither sea ice nor open water. When he sailed away for his period of exile, he set a north-west course, one that describes the shortest distance between Iceland and Greenland. However, under normal circumstances, you can't see Greenland from Iceland, even if you are standing in pastures a hundred metres above the sea. Nor are there any fishing grounds northwest of the island that might have provided an opportunity for a sighting. Sailors would have to have travelled more than a hundred and fifty kilometres towards Greenland before sighting land.

The only way an image of Greenland could be seen in Iceland is if the atmosphere over that part of the North Atlantic settled into a temperature inversion, warm over cold, creating the same curving paths for light that created the image of Toronto in Buffalo, or allowed Captain Bartlett to see five hundred kilometres across the North Atlantic. Did a mirage create the legend that Erik chased? Waldemar Lehn argues that there must have been some powerful reason for Erik to take that particular course, because it was not the best from a navigational point of view. He was sailing into both currents and winds and because Viking ships were incapable of tacking, his crew might have had to row a good part of the way. Sailing more to the west would have allowed them to pick up favourable currents and easterlies and they likely would have arrived at their ultimate destination in southern Greenland with a minimum of effort.

Stranger still, mirages might have made navigating the North Atlantic more frightening. It wasn't the voyage itself: even if Norse navigational knowledge was skimpy, their ocean-going vessels were anything but primitive, capable of transporting several dozen crew manning six-metre oars under a huge woollen sail. The real concern was that the oceans were said to contain weird natural phenomena and monstrous creatures. Two of the most prominent of these might have been mirages.

One of these was the whirlpool, called the "*hafgerdingar*" by the medieval Norse; a "sea-fence" or "sea-hedge." These are described in *The King's Mirror*, an anonymous collection of information about Ireland, Iceland and Greenland written in the thirteenth century: "These hedge in the entire sea, so that no opening can be seen anywhere; they are higher than lofty mountains and resemble steep, overhanging cliffs." *The King's Mirror* describes them as occurring in three successive waves, leading one researcher to suggest these were tsunami created by sea quakes.

Lehn is skeptical of quakes, arguing that their rarity, combined with the small numbers of sailors travelling between Iceland and Europe, would have made the likelihood of experiencing a sea quake (let alone surviving it to tell the tale) vanishingly small. More likely, he thinks, are those powerful mirages where the horizon seems to rise up on all sides. Lehn and his colleagues have experienced these, both in the Arctic and around the shores of Lake Manitoba. He has found that the visual brain exaggerates the true degree of light bending, so that a slight effect causes, as he put it, "the waters to appear to be poised as if to engulf the observer." The effect would be incredibly dramatic: walls of water would surround a ship on all sides, and sailors would feel as if they were clinging desperately to those walls, about to plunge into the maw of a giant whirlpool.

In fact, in some concepts of the world at the time, whirlpools played an essential role. The flat earth was said to be surrounded by an ocean on all sides, an ocean pockmarked with whirlpools which served to return to the land water that had flowed into the ocean through the world's streams and rivers. Whether the people plying the waters around Iceland knew about or believed this scenario, had they experienced the vivid mirage of a whirlpool, they would surely have had good reason to believe in its existence.

But whirlpools were not the only thing for Norse sailors to fear— there were the strange beasts of the sea as well. *The King's Mirror* is a thorough guide to the natural history of the time. It lists several different kinds of whales in Icelandic waters, including several species recognizable today, like sperm, humpback and narwhal. It also describes whales, walruses, polar bears and seals in and around Greenland. It is unlikely that any of these would have been terrifying to hardbitten sailors. But the merman might have been.

The King's Mirror claims that the merman lives in the waters around Greenland: "This monster is tall and of great size and rises straight out of the water. It appears to have shoulders, neck and head, eyes and mouth, and nose and chin like those of a human being; but above the eyes and eyebrows it looks more like a man with a peaked helmet on his head. It has shoulders like a man's but no hands. Its body apparently grows narrower from the shoulders

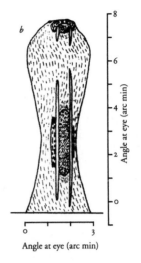

On the left, a walrus. On the right, that same walrus seen through the distorting lens of the atmosphere. Is that what Norse sailors described as the mysterious and forbidding merman?

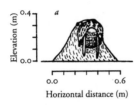

down, so that the lower down it has been observed, the more slender it has seemed to be. But no one has ever seen how the lower end is shaped, whether it terminates in a fin like a fish or is pointed like a pole. The form of this prodigy has, therefore, looked much like an icicle . . . Whenever the monster has shown itself, men have always been sure that a storm would follow."

Embedded as it is among accurate descriptions of sea life we are familiar with today, this seems bizarre beyond belief. A huge sea animal that tapers from top to bottom? This is a case where —for once—the "optics of the situation" is literally accurate. It is possible to envision atmospheric circumstances that could produce mirages not unlike those described in *The King's Mirror*.

In a report in the journal *Nature* in January 1981, Lehn and his colleague Irmgard Schroeder showed how the image of a walrus with just its head and shoulders out of the water could, with the right temperature, be transformed into a grotesque, bowling-pin-shaped creature with a squashed walrus head on top and a second stretched-out image of the muzzle and tusks covering most of the body. It isn't a perfect replica of the merman, especially because it tapers in the middle, expanding again at its base, rather than tapering to nothing as *The King's Mirror* describes. However, the image is of a mighty fearsome and unusual creature.

Lehn and Schroeder were able to create the image by processing a photograph of a walrus in ways that mimicked natural atmospheric layering. The walrus was placed (theoretically) about three-quarters of a kilometre away in cold water. The layer of air above the water varied in temperature from about one degree Celsius at water level to eleven degrees two metres up. But the increase wasn't uniform: at a point about a metre and a half above the water the temperature increased rapidly over a short distance. This critical point, called the thermocline, plays a key role in the generation of the mirage. If the walrus is viewed from a vantage point just below the thermocline, at a height of just less than a metre and a half, the mirage is dramatic.

In their *Nature* paper Lehn and Schroeder produced a photograph taken on May 2, 1980, of a boulder on Lake Winnipeg. The air temperature was twenty-eight Celsius, but there was still ice on the lake, perfect conditions for establishing a temperature inversion. The innocuous boulder, a little more than a kilometre away, turned into a strange merman-like figure.

We will never know whether such Arctic mirages were guideposts for Erik the Red or created the fantastic creatures the Norse supposed existed in the oceans of Greenland and Iceland. If they were, it is a reminder that much was still mysterious then; the wildlife of the oceans hadn't yet been catalogued and mysterious forces were still believed to be operating. It wasn't just that the visual brain of a Norse sailor could be fooled—it was that the rest of his brain was then free to spin a tale around that perception. Of those two steps, only the first happens today. We live, fortunately or not, in a world where myth-making is much more rigidly constrained.

Sane in an Insane World

On January 19, 1973, *Science* magazine published a study that would create a sensation in the psychiatric community in North America. It was called "On Being Sane in Insane Places," and was written by David Rosenhan, a professor of law and psychology at Stanford University. Rosenhan was asking a profound question: Can we really tell the difference between sanity and insanity? He performed a dramatic experiment to answer that question.

Rosenhan persuaded eight volunteers from different walks of life, including a painter, a housewife, three psychologists, even a psychiatrist, to become "pseudopatients." They were to seek admission to a psychiatric hospital, on the grounds that they had been hearing voices. Their testimony in each case was exactly the same: they reported to hospital personnel that they had been hearing voices saying three words (and no others): "empty," "hollow" and "thud." Those words had been chosen to suggest that the pseudo-patient was experiencing some sort of existential crisis; they were also chosen because there was not a single example of a psychosis of that type on record. Apparently, although we may agonize over the meaning of life (or its lack), it doesn't drive us crazy.

Beyond claiming these symptoms and falsifying their names and occupations, the pseudopatients were completely truthful. All questions about family background, relationships and any potential emotional disturbances were answered straightforwardly. In other words, apart from claiming that they were hearing voices saying those three words, the patients acted completely sane, for sane they were.

Apparently the pseudopatients were most concerned about the possibility they would be found out immediately upon arrival and greatly embarrassed as a result. It turned out their fears were unfounded. All were admitted to hospital. In fact, some did double duty, with the result that the eight pseudopatients were admitted to twelve hospitals in all. Once there, they no longer mentioned the hallucinations, and if asked directly, claimed that they had stopped. So from the moment of admission on, with the apparent exception of some nervousness, they acted exactly as they would have on the outside. Rosenhan's question was: How long would it take for the personnel in the psychiatric hospital to realize that these were fakers, sane individuals among the insane?

It took some time to answer that question fully: one of the pseudopatients didn't get out of the hospital for fifty-two days! The shortest stay was five days; the average was nineteen. All but one of the pseudopatients were diagnosed upon admission as "schizophrenic"; all but one left with the label "schizophrenia, in remission." (The other pseudopatient was diagnosed as manic-depressive.) Rosenhan took these labels to indicate that the hospital personnel never did recognize that the pseudopatients were sane. This dramatic demonstration that days or weeks of observation of sane people in psychiatric hospitals failed to make apparent their sanity convinced him that the definition of sanity depends totally on the context. If you're sane but in an insane place, even experts think you too are insane. Rosenhan then argued that psychiatric diagnosis itself must be unreliable if it is incapable of separating the individual and his/her symptoms from the setting. Is insanity in the mind of the observer or the observed?

Anticipating protests that because accuracy can never be 100 per cent, it is better for psychiatric experts to diagnose a healthy person sick than a sick person healthy, Rosenhan went on to design a second experiment. In this case he warned a teaching and research hospital that pseudopatients were coming, one or more over the ensuing three months. In actual fact, only one pseudopatient had been recruited, and he had to back out at the last minute. But that didn't ruin Rosenhan's experiment. Of the 193 real patients who were admitted over that three-month period, many were suspected of faking. One psychiatrist considered 23 of them suspect—another staff member had serious doubts about 41. So failure swung in both directions, more evidence for Rosenhan that psychiatric diagnosis was fatally flawed.

Obviously research like this that struck at the roots of psychiatry wasn't going to go unchallenged, as the response to Rosenhan soon showed. But in addition to the questions raised about the reality of diagnosis, the experiment also produced some uncomfortable revelations about life behind the hospital doors.

First, an irony. The sanity of the pseudopatients may have escaped the notice of the experts, but the patients weren't fooled. In three hospitals, 35 patients (out of a total of 118) were suspicious about the true identity of the pseudopatients: "You're not crazy. You're a journalist, or a professor. You're checking up on the hospital." (These suspicions were based on the fact that Rosenhan's volunteers were writing down their observations of daily life in the hospital.) Obviously it could be argued that patients who belonged in the hospital might have a skewed impression of sanity (for instance, we have no idea whether they thought any of the real patients were faking it too) but Rosenhan's colleagues were pretty sure that wasn't the case, that in fact the inmates detected signs of sanity where the staff didn't.

Perhaps because patients had no access to the pseudopatients' medical records, their perceptions weren't coloured by the diagnosis of "schizophrenia." However, it became apparent that the same unbiased approach wasn't shared by the staff. Rosenhan argued that the label—a label which he had shown to be inappropriate—coloured the behaviour of his pseudopatients in the eyes of the hospital staff, making it much more difficult for sanity to emerge. For example, one pseudopatient related a story of changing relationships with his parents. As a young child he had been closer to his mother, but beginning in adolescence, he befriended his father and grew more distant from his mother. Here is a portion of his case history as recorded by hospital staff:

> This 39-year-old white male . . . manifests a long history of considerable ambivalence in close relationships, which begins in early childhood. A warm relationship with his mother cools during his adolescence. A distant relationship to his father is described as becoming very intense. Affective stability is absent . . . and while he says that he has several good friends, one senses considerable ambivalence embedded in those relationships . . .

Would that description have been the same if the writer had assumed the patient was "normal"? There were even better examples. How about this comment, recorded daily by a nurse watching a pseudopatient write down his observations in a notebook: "Patient engages in writing behavior." The assumption of illness coloured observations of real as well as pseudopatients. There was the psychiatrist who told medical residents that a group of patients standing outside the cafeteria a half-hour before lunch were exhibiting behaviour characteristic of the "oral-acquisitive" nature of their syndrome. Rosenhan figured a better description was that they were bored and hungry.

The Rosenhan experiment went on to document a variety of ills in the hospital, ranging from avoidance of patients to abusive treatment of them. Pill-giving seemed to be the main activity of the staff—2100 of them were given to the pseudopatients in all, of which 2098 were pocketed or flushed down the toilet. What pseudopatients and their real counterparts did with their medications went unnoticed by the staff. And these weren't uniformly run-down underfunded psychiatric hospitals; some were old, some new; some understaffed, some not. All but one were publicly funded.

Even today, reading "On Being Sane in Insane Places" is unsettling—it fits so well with our fears and suspicions of mental illness; it makes concrete the nightmare of how we could be labelled, defined and put away on the basis of an unjustified diagnosis. It characterizes the staff in these hospitals as uncaring, unobservant, with personalities ranging from violent to austere. Not surprisingly, it attracted all kinds of response from professionals in the field, most of it denying Rosenhan's main claim that psychiatric diagnosis was virtually useless.

There was some agreement: Rosenhan and many of those who commented on his experiment agreed that accepting the pseudopatients into hospital was the right thing to do—the patients seemed to have been suffering from hallucinations and appeared nervous. Diagnosis aside, the humane thing seemed to have been to admit them. However, that was practically the only point of agreement. Many critics argued that what had gone on in the twelve unnamed hospitals would never have happened at theirs. They raised questions like, "How could the staff not have been curious about the odd symptoms reported, symptoms that had never before been mentioned in the psychiatric literature? How could the staff not have interviewed the families of the pseudo-patients? How on earth could anyone diagnose schizophrenia on the basis of reported auditory hallucinations alone . . . that would never happen in my hospital." I'm going to leave aside arguments of this kind that suggest the hospitals in the experiment were somehow unusual and focus instead on the problems in the business of psychiatry highlighted by this experiment, not the institutions in which it is practised: the diagnosis of schizophrenia, the fact that the patients were never discovered to have been faking, the length of time they spent in hospital and the fact that they were released as "schizophrenic, in remission."

Some of the most striking criticisms attacked what seemed to be the most uncontentious part of the study, the diagnosis. At first glance Rosenhan seems to have been completely justified in saying the most worrying part of the experiment was the willingness of the hospital personnel to diagnose pseudopatients as schizophrenic based on a single symptom: that they had been hearing voices. In fact it wasn't even a symptom, it was a report of a symptom. In all other respects except for a little nervousness, the pseudopatients acted normally. Rosenhan was shocked that the leap was made so easily from this, the slimmest of evidence, to a diagnosis that was never reversed, only amended.

Many of his critics (I'll omit the long list of names here—the collection of original commentaries is listed in the bibliography) admitted that a definitive diagnosis of schizophrenia was a stretch, and that something like a set of hypothetical diagnoses, ranked in order of likelihood, would have been more sensible. Yet others argued that in the absence of other information, with no evidence of unusual recent stresses or mood disturbances, schizophrenia is the best available explanation for hearing voices: a plausible diagnosis, especially if other possible causes, like alcohol withdrawal, use of hallucinogenic drugs or the presence of some kind of brain damage, are ruled out. It was impossible to know if these had been addressed because Rosenhan never revealed the contents of the admission interviews, a fact that not only left unclear how thorough the interview had been, but also left critics free to suggest that much more might have been said by the pseudopatient in the interview that would suggest mental illness.

Several critics took issue with Rosenhan's conclusion that psychiatric diagnosis has been revealed to be a bogus exercise. In fact they drew exactly opposite conclusions, claiming that the fact the pseudopatients were faking invalidated the whole approach and that the only justifiable conclusion from the experiment was that people who aren't suspecting fakers cannot tell the difference between the insane and the "sane-feigning-insanity." One said simply that what Rosenhan had shown with the experiment was that 1) sane people who fake a history of hallucinations but behave normally in hospital are not considered abnormal, and 2) if a psychiatric patient behaves normally in the hospital, that person is not suspected of having faked his/her symptoms, two not very startling conclusions. The issue of fakery was made most vivid by the suggestion that if someone were to drink a quart of blood, then run to an emergency room and vomit the blood, the staff would likely diagnose that person as having a peptic ulcer. Could it then be argued that medical science doesn't know what it's doing when it comes to diagnosing ulcers?

There were alternative experiments suggested: repeat the Rosenhan experiment but have the pseudopatients feign nothing, simply present themselves to the hospital, act normally, tell their life story exactly as it was, then ask if they could be admitted as patients.

The other side of that experimental coin would be to take actual hospital-bound schizophrenics, transfer them to new hospitals and see if those institutions would turn them away as sane. Neither seems very likely. Another suggestion was to mix together twenty psychotic people and twenty non-psychotic and see if psychiatrists could tell them apart. When you consider those alternatives, the Rosenhan experiment loses some of its power.

Others scoffed at Rosenhan's claim that the pseudopatients acted completely normally except for the report of having heard voices. After all, they sought admission to the hospital—that is obviously extremely unusual behaviour for someone who is not mentally ill; it's also an admission that the "voices" were particularly troubling. The pseudopatients' note-taking prompted some skeptics to note that this is common behaviour among patients in psychiatric hospitals and could easily have made the pseudopatients resemble "compulsive paranoid patients." This touches on one of Rosenhan's main complaints: that the diagnosis resides with the place, the context, not in the person. Writing is okay unless you're being observed by a nurse who believes you're disturbed, then it's suddenly symptomatic.

But this point could be argued. After all, the pseudopatients were assumed to be there for a good reason (they had apparently been hallucinating, they asked for admission) and so staff should be expected to make note of anything that might shed light on their "condition." In fact one commentator pointed out that recording "patient engages in writing behaviour" didn't imply the nurse thought that that was pathological. Good staff note everything. His only complaint was that several words had been used where one would have been sufficient.

Some even claimed that if the patients had indeed been acting completely normally, they should have admitted to the staff that the whole thing was a trick:

Look, I am a normal person who tried to see if I could get into the hospital by behaving in a crazy way and saying crazy things. It worked and I was admitted to the hospital, but now I would like to be discharged from the hospital.

Rosenhan thought this behaviour would likely have had an effect exactly opposite to that desired, that it would have been "precisely the insane thing to do." In fact, he and his colleagues asked real patients what they would do to get out, and they never advised that direct approach. One said, "Don't tell them you're well. They won't believe you. Tell them you're sick, but getting better. That's called insight, and they'll discharge you."

Several critics took issue with the claim that releasing the patients with the label "schizophrenia, in remission" was a bad thing. One wrote that as far as he was concerned, this meant that the patients had been experiencing hallucinations, but in their time in hospital, weren't. So they had been psychotic, now they weren't, so they were in remission. His conclusion? In twelve cases out of twelve, the diagnosis was correct, an accuracy rate of 100 per cent. Another pointed out that describing a released patient as being "in remission" was a highly unusual occurrence. In the hospital in which he worked, only one of the last hundred schizophrenics discharged had been labelled "in remission." Other hospitals reported similar rarity, leading him to the conclusion that, rather than lamenting the diagnosis, "we must marvel that . . . psychiatrists acted so rationally so as to use at discharge the category 'in remission' . . ."

Rosenhan maintained that "in remission" meant that the psychiatrists couldn't tell the difference between sane and insane. This worried him, not only because it seemed a fatal flaw in psychiatric diagnosis, but also as a route to assigning damaging labels to innocent people :

Consider two people who show no evidence of psycho-pathology. One is called sane and the other is called paranoid schizophrenic, in remission. Are both characterisations synonymous? Of course not. Would it matter to you if on one occasion you were designated normal, and on the other you were called psychotic, in remission, with both designations arising from the identical behaviour? Of course it would matter.

But even this persuasive appeal was dissected. If there were two such individuals, would their future course be identical? Re-admissions for schizophrenia, at least in the early 1970s, were double the number of first admissions. Within the first year, a third of schizophrenics re-entered hospital. It would have been foolish, the argument went, to pretend that a schizophrenic released from hospital was in the same position as a patient whose broken arm had mended. One was still at risk, and in the minds of many of Rosenhan's critics, that simple phrase "in remission" was an acknowledgement of an unfortunate but real situation.

What about the fact that it took so long—as much as fifty-two days in one case—for the pseudopatients to be released? Critics posed the alternative question: What would be the minimum time these patients could have spent in hospital, given that they were believed to have been suffering from hallucinations when they were admitted? One gave the following sample interview as an example:

Diagnostician: Tell me what is wrong with you.
Patient: I often hear voices that say, "hollow," "empty" and "thud."
Diagnostician: Are you hearing the voices now?
Patient: No.
Diagnostician: Good. There is nothing wrong with you.
Go home.

The writer freely admitted this was an extreme example but insisted it highlighted a crucial problem: When do you diagnose someone as no longer hallucinating? Imagine you're betting on the toss of a coin, and after winning and losing, winning and losing in about a 50:50 ratio, you start to lose flip after flip. How long would it take you to realize that the game had been rigged? A long time if you have no reason to suspect that it was. How long would it take a psychiatric hospital professional to realize that someone who had been admitted apparently suffering from hallucinations was acting completely normal?

It isn't uncommon to see symptoms of mental illness fade in the hospital setting, in the absence of triggers that might exist at the patient's home, work or simply in the street. More than one critic pointed out that life in a psychiatric hospital isn't the *One Flew over the Cuckoo's Nest* stereotype—patients can appear calm, somewhat subdued, anything but unusual. In that sense the pseudo-patients may not have appeared so different from their real counterparts. So "normal" behaviour in hospital doesn't necessarily mean that the person is completely free of whatever brought him to the hospital in the first place. It therefore shouldn't be surprising, the critics argued, that hospital staff would be cautious in reaching the conclusion that a patient could be safely released.

Obviously psychiatrists didn't take the Rosenhan experiment lightly. But to be fair, while they all had reservations about the sweeping statements he had made about the value of diagnosis and the inability of psychiatry to distinguish between sane and insane, many admitted the value of some of his claims. For instance, given that the original diagnosis should have been made with caution, and given that the patients were all discharged because they were showing no signs of schizophrenia, why was it that no one admitted that the original diagnosis might have been in error?

There was also agreement that conditions in the hospitals could be dramatically improved. Rosenhan had claimed that staff seemed to do their best to avoid patients, not communicate with them; that their attitudes towards patients were a combination of "fear, hostility, aloofness, suspicion and dread" and that the experience of being hospitalized was "depersonalizing." While some critics contended that such was not the case in their institution, most doubted that the conditions could have been as bad as they were painted or simply ignored this part of "On Being Sane in Insane Places." It might have been because there was already acknowledgement that conditions weren't ideal in many psychiatric hospitals, but I think the real reason is that mental health professionals were, if not threatened, annoyed by Rosenhan's claims that diagnosis was an inaccurate, unhelpful and long-lasting labelling of patients. That claim undermines the whole profession, and it met with the response you'd expect.

Most of the furor had died out by the end of the 1970s. What would happen if Rosenhan's experiment were to be conducted today? Dr. Donald Wasylenki, a psychiatrist at the University of Toronto, argues that the problem today wouldn't be so much admitting pseudopatients, but failing to admit real ones. It's not just that private health insurance companies would be demanding more say in a patient's admission; he points out that in the early 1970s there were lots of beds in psychiatric hospitals, and many patients in so-called "custodial" care. Now after years of deinstitutionalization, thousands of beds have been lost.

Dr. Wasylenki also feels that diagnosis is much more precise now with the widespread use of the DSM-IV, the *Diagnostic and Statistical Manual of Mental Disorders*, Fourth Edition, an 886-page tome published by the American Psychiatric Association. Dr. Michael Bagby of the Clarke Institute of Psychiatry in Toronto agrees that DSM-IV is a much more heavyweight (literally) guide than the second edition, which was in use at the time of Rosenhan's experiment.

These comments wouldn't comfort those critics today who see the same weaknesses in psychiatry that Rosenhan claimed. But that is part of a much larger debate. The Rosenhan story itself is complete, and we can all take from it what we want. And I'm ambivalent: some of the critiques are eloquent and to the point, and they weaken Rosenhan's claims. But I (and maybe I was chronically infected by the anti-psychiatric sentiments of the time) can't help thinking that he uncovered more truths than most professionals were willing to admit.

The Barmaid's Brain

———
———

Even if one day scientists completely understand the wiring and chemistry of the human brain, it will still be difficult not to be amazed by an organ that can memorize the lyrics to all Spice Girls' songs after one hearing or conjure up equations describing the origin of the universe. Even more amazing is that the same brain can do both. Notwithstanding Deep Blue's defeat of Gary Kasparov, the brain still reigns supreme over the most powerful computers. But its power and versatility is even more remarkable considering that it is built from the same stuff as all the other brains on the planet. The human brain has been assembled, piece by piece, over millions of years. There was no point in human evolution when our brains were suddenly and completely remodelled, never a dramatic evolutionary event to install the biological equivalent of a Pentium II.

As a result of this construction of a mansion out of a tool shed, we share much of the basic wiring in our brains with other creatures, some of them considerably less brainy than we are. The feeding behaviour of the leech employs the same neurotransmitter—serotonin—that in the human brain is boosted by Prozac; we have emotional centres similar to those in rodents, and even the cell-by-cell circuits responsible for paying *attention* and having *intention* have been traced in the monkey brain.

For reasons that aren't yet clear, however, the frontal lobes of the human brain, the centres of higher reasoning, grew dramatically over the last half-a-million years, leaving us unusually brainy. Judging by its shape and size, and the marks left on the inside of fossil skulls, the human brain had the same organization and presumably the same potential a hundred thousand years ago as it does today.

We have a brain whose structures and capabilities are largely foreshadowed in other animals, but at the same time possess added brain power that is uniquely ours. That power is exemplified not so much by sheer computing ability as by flexibility and adaptability. For instance, our brains are capable of mental feats that couldn't have played a role in our ancestors' lives hundreds of thousands of years ago. After all, written language and mathematics have only appeared in the last few thousand years; our abilities to do both are somehow piggybacked onto neural circuit boards that are, at the very least, 100,000 years old. But it is not just that our brains have shown the ability to take on new roles over millennia—they can also adapt within a lifetime. For instance, there is an inordinately large amount of space within a professional musician's brain dedicated to the perception of musical notes. Sometimes our brains adapt to the demands of their environment in that way, by dedicating additional neural space to specific activities; sometimes the reverse happens, and the brain abandons certain mental skills.

Psychologists and neuroscientists design elaborate experiments in an effort to understand that balance between being captive to our evolutionary past and having the ability to transcend it. But the psychology lab isn't the only place where such insights can be gained—sometimes the lab of everyday life is better. And one such lab is the cocktail lounge. Being a waiter in an establishment serving alcoholic beverages is, in evolutionary terms, a recent challenge for the human brain. The earliest evidence for brewing beer dates back a few thousand years, while lounges could only have arisen in towns, which first appeared around the same time. So the mental skills necessary to serve a martini or Scotch-on-the-rocks must, like writing and math, be adapted to the brain architecture already present. This may explain why the brains of such people exhibit both extraordinary capability and unusual inability—in the same head.

There are two outstanding experiments in this admittedly rare (or should I say exclusive) field of study, but before I detail them, a note of political correctness. I have no problem with the use of the generic term waiter to refer to either sex. However, the two studies I'm about to report focus on females and use the term waitresses (both were published before the non-sexist generic was in common use), and to make things more consistent and easier on you and me I'm going to retain that terminology.

The first study concerns the women who serve beer in Munich during Oktoberfest. I had the privilege, many years ago, to watch these women at work, both in the famed Hofbrauhaus and the Mathaser Bier Stadt (though not on the same night). They are truly remarkable for their speed and strength, bustling from table to table carrying at least five one-litre mugs of beer in each hand. In the Bier Stadt, at least when I was there, these mugs were heavy glass; in the Hofbrauhaus the waitresses had a slightly easier time with the traditional ceramic vessels. But these places were crowded, the waitresses moving quickly, and obviously there was a premium placed on getting the beer from the bar to the table without spilling it. Curiously enough these hard-won physical skills have caused the waitresses to lose some mental acuity.

Child psychologist Jean Piaget made famous the demonstration that the understanding of liquids in containers is something that develops gradually. Young children have not yet developed a sound concept of liquid volume and can watch as you pour water from a short squat glass to a tall narrow one and then assert with confidence that the tall glass has more water in it than the short one did. It's not the volume they're paying attention to, but the height. Most of us as adults have no problem figuring that one out, but at least 40 per cent of us are stumped by another Piagetian puzzle: if you're holding a glass half-full of water, and you tilt the glass until the water starts to pour out, is the water level in the glass now horizontal, just above horizontal, or just below?

This question is much better adapted to pencil and paper. You are presented with a simple line drawing of a tilted glass over a table (to establish horizontality) and asked to draw the water surface inside the glass. A consistent forty per cent of adults gets this test wrong, asserting that the water is tilted either to the left or right. In fact it is exactly horizontal—gravity wouldn't have it any other way. What is curious about the large number of people who fail the test is that some of them know that water will always remain horizontal, but draw the water line incorrectly anyway, while others who haven't a clue what should happen nonetheless draw the water level accurately. Clearly there is some sort of disconnection here between knowing and drawing.

Experiments by Dennis Proffitt at the University of Virginia explored this psychological blind spot, ultimately taking it right into the Munich beer halls. Profitt first tested the idea that pencil and paper tests might be so unrealistic that they are giving the wrong impression of what is in people's minds. In these tests any line representing a water level that departs from the horizontal by more than five degrees is classified as inaccurate. That might not sound like much of a disparity, but you know immediately when you look at it that it is clearly not horizontal. The question is whether that is an accurate representation of what's in the person's mind or an artifact of drawing lines on paper.

To answer this question, Proffitt presented volunteers with a video of a glass jug containing a dark liquid held at various angles. By tilting both jug and camera he was able to produce both realistic and impossible images. Despite the realism of this imagery as opposed to line drawings, his subjects still made all kinds of mistakes. Proffitt even admitted that many of the "impossible" orientations of the liquid in the jug looked perfectly reasonable to him.

He then hypothesized that the problem was one of context. When faced with this problem you can compare the surface of the liquid to one of two things: the container or the tabletop. If you're relating the liquid to the container you're in trouble, because the container is tilted and there is no simple way of comparing the surface of the liquid to the horizontal. On the other hand, comparing the liquid to the tabletop and making them parallel, is a simple, effective, but easily overlooked, solution. As Proffitt pointed out: "Hold a full drink, and observe the liquid's surface as the drink is brought to the mouth. The orientation of the surface relative to the lip of the container demands attention, whereas an awareness that the liquid's surface orientation is horizontal does not come spontaneously to mind." To say the least.

To test the hypothesis, Proffitt and his colleague Ellen McAfee designed a new version of the traditional pencil-and-paper test. They presented volunteers with photos of a darkened liquid in three different kinds of containers: a traditional lab-style beaker and two circular containers, a goldfish bowl-shaped beaker and a petri dish. (The latter looked a little like a transparent hockey puck.) The idea here was to draw attention away from features like the lip. The photos themselves were circular and were mounted on a turntable that allowed the subject to rotate them until he/she felt the level of the liquid was oriented the way it would be in real life (remember that these are only photos, so the liquid isn't sloshing around as the turntable turns). There were no cues to horizontal in the photographs themselves. Another novelty in this experiment was that the liquid and its container were fixed relative to each other—tilting the photo didn't bring the liquid closer to or further from the lip—in fact, the petri dish had no lip. So subjects were free to bring the level of the liquid into position without distraction.

The results of this experiment were remarkable to say the least. Of twenty-four people who had failed the pencil-and-paper test, nineteen were correct with the beaker photographs and every single one of them judged the liquid level in the petri dish correctly. The results were solid support for Proffitt's argument that the position of the liquid *vis-à-vis* the lip of the container is a distracting influence, and once that influence is removed, most people can judge the level of the liquid accurately.

At least for a moment or two. When Proffitt immediately gave these same subjects the traditional pencil-and-paper version of the test, only about a quarter of them retained their new sophistication; the rest made the same mistakes they always had. Proffitt had already noted that females tend to be less accurate than males in the water-level test and similarly older people did less well than younger. But now a question arose: Would experience with liquid-filled containers help or hinder performance on this test? Obviously it was time to take the problem into a natural setting, and there is no better place than Munich, and no better time than Oktoberfest.

Here then was the line-up for the crucial experiment. Six different groups of subjects took part: male and female students at the Ludwig-Maximilians-Universitat Munchen, stay-at-home mothers, male bartenders, male bus drivers and finally, waitresses who served beer at a local brewery during Oktoberfest. The waitresses' jobs were as demanding as I remembered them: hauling five one-litre containers of beer in each hand through a crowded hall. All the subjects were given the pencil-and-paper version of the water-level test, and asked to draw the level in a glass that was apparently being poured into a bowl on a table.

The students did best, females scoring 65 per cent correct and males 80 per cent. That male-female disparity was exaggerated in the housewives and bus drivers, who scored 30 per cent and 75 per cent respectively. But the real shock came in the results for the bartenders and barmaids, whose jobs involve first-hand, daily experience with liquids in containers. Bartenders scored a dismal 40 per cent (much lower than the other males) and the waitresses, a mere 25 per cent. That wasn't all: the more experienced the waitress, the larger the error in her estimate. In addition, during the post-experimental debriefing both bartenders and waitresses expressed surprise at the correct answer, sometimes remaining unconvinced until the experiment was actually demonstrated.

Dennis Proffitt's explanation of these results was straightforward. The biggest demand on an Oktoberfest waitress is to get the drinks to the tables without spilling them. The best way to ensure that is to pay attention, not to the level of the beer with respect to the ground (that would guarantee short-term employment) but with the lip of the glass. As Proffitt showed in previous experiments, people who focus on that relationship do poorly in the water-level test, as did the Oktoberfest barmaids.

This experiment illustrates nicely the adaptability of our brains. We all begin life with roughly the same brain wiring, the "default" human wiring, meaning in this case the same ability to judge the orientation of liquids. But we are adaptable, and while a contractor might know that the level of beer in a glass perched on a beam should be horizontal, that is completely irrelevant to a waitress. As long as the beer stays in the glass, she's happy. I bet if you evaluated the waitress's neural motor programmes for maintaining balance and minimizing spillage, you'd find she would be at the top of the class.

There is a postscript to this experiment: females in general do worse on the judgement of liquid levels than do males. I'm sure that some psychologists would be tempted to explain this difference in terms of "evolutionary psychology," the idea that because we were hunter-gatherers in the final phases of the evolution of the modern brain, male-female role differences were preserved in its structure and wiring. But while the observed superiority of females to remember objects on a table might relate to their role as knowledgeable gatherers of patchy food resources scattered over the landscape, it's a lot trickier to explain the ability to judge whether a liquid is horizontal or not. Dennis Proffitt, the man who designed the experiments, is in this case reluctant to offer any evolutionary explanation.

So under the pressures of serving in the beerhalls of Munich, barmaid's brains have lost some of their perceptual abilities. But there is a different experiment with waitresses that illustrates how, if you take a different point of view, their mental abilities appear to be not just extraordinary, but unexplainably so.

This experiment was conducted by Henry Bennett at the University of California, Davis, in the early 1980s. (I outlined the experiment in *The Burning House*.) At the time Bennett was interested in memory research and was convinced that despite the all-too-apparent limits to human memory capacity, cocktail waitresses somehow managed to escape those bounds. He pointed out that waitresses must remember (without writing down) drink orders from several different places in a bar—as many as fifteen in a single trip—rearrange those orders into what's called the "calling order" (a sequence that allows the bartender to make the drinks as fast as possible), hold a conversation, then place each drink when ready in front of the appropriate customer. (Nothing is worse than arriving at a table and saying, "Who's having the . . . ?") Waitresses obviously learn these memory skills, but Bennett wanted to measure just how superior they are.

His experiment was peculiar, to say the least. In order to test the ability of waitresses to remember drink orders, he devised a miniature cocktail lounge. This included two tables fourteen centimetres across, surrounded by little chairs two centimetres high. The "customers" were dolls dressed and made up to look either like men or women—they were just under ten centimetres tall. The drinks (thirty-three in all) were rubber stoppers for laboratory glassware with flags stuck in them bearing the names of the drinks. Voices recorded on a cassette tape player uttered the orders: "Bring me a margarita . . . (two second pause) . . . "I'll have a Budweiser."

Professional waitresses and students were tested for their ability to remember the drink orders in this miniature lounge. A subject would watch as first seven customers, then eleven and finally fifteen ordered drinks. As each voice ordered, Bennett would wiggle the appropriate doll so that the subject knew who was ordering what. Following the orders there was a two-minute break while the subject was asked questions, then the subject was asked to distribute the drinks to the customers.

Here's how they did. Waitresses averaged 90 per cent accuracy on their orders; students 77 per cent. Those results include the relatively easy seven-order part of the experiment; when it came to remembering fifteen drinks, the waitresses were head and shoulders above the students: 86 per cent to 68 per cent. Six waitresses (out of a total of forty) placed all thirty-three drinks correctly, and two of those had heard the drink orders out of sequence, that is, in random order around each table. These two stellar waitresses also excelled in speed—it took them a little more than two seconds per drink to distribute the drinks to the correct customers around the two tables. Bennett figured that was about as fast as was physically possible.

How could these waitresses accomplish these amazing feats of memory? It's worth noting that random ordering, as opposed to taking drink orders in sequence around the table, makes it impossible to use any sort of positioning cues for remembering. As Bennett noted, this controlled experiment in a peaceful setting didn't exactly mirror the real-life ambience of a cocktail lounge, where the memory feats are even more remarkable. Compare this to the well-known psychological principle that most of us can only manage to hold about seven "chunks" of information in our short-term memory at any time—a good example is keeping a phone number in mind as you walk from the phone book to the phone. There are individuals who after training diligently can remember far more, but they have made a dedicated effort to improve their memories, while the waitresses seemed to have acquired the skill on the job. The question was, how do they do it?

Most of the waitresses claimed that their memories were better when they were busiest and under greater pressure to remember. One of them described it as "then I'm in the flow," and several told some truly incredible stories. One waitress remembered serving a party of twenty-five, all of whom had separate bar and food tabs; she was able to prepare all fifty bills without writing anything down. Another waitress—with only three years' experience—told of working New Year's Eve when the two other waitresses scheduled to work called in "sick." She had one hundred and fifty customers that night:

By the end of the night I knew what every customer was drinking. I'd just stand by the bar, looking for hands, and give the bartender the order. Then I'd take the drinks over to the table. I really don't know how I did it.

These remarkable performances are only partly explainable by the techniques waitresses use to aid their memories. Sometimes they try to guess what customers are going to order before they actually do—even if they're wrong, their attention is focused completely on the order. At least three waitresses testified that after a while, "customers start looking like drinks." That sense can be strengthened by connecting appearance with order: the white guy with the black Russian, the flushed woman drinking the strawberry Daiquiri. Sometimes a waitress remembers a face; sometimes she repeats the order silently to herself. On busy nights she may benefit from the familiarity of the lounge to be able to use the age-old method of attaching the drink to the place: "a dark beer in the far corner, a martini next door and white wine under the lamp-shade . . ." in the same way that you can remember a shopping list by mentally placing the food items in various places in your house.

The one clear message was that writing down orders is inefficient and not very popular. Most of the time waitresses cannot really explain how they remember. And Bennett came to the astonishing conclusion that there might not be an upper limit on the number of drinks a good waitress can remember.

The motivation to remember was agreed upon: the better the accuracy, the more drink orders and the higher the tips. Customers seemed flattered when waitresses remembered their order. I don't know how good the tipping is during Oktoberfest, but I would be surprised if speed (and lack of spillage) in that case were not motivated by the same pecuniary factors.

So we have two experiments, two sides of the coin. One shows professional waitresses to be inferior in one narrow perceptual ability, the other demonstrates a dramatic superiority in memory.

But in one important sense these experiments are anything but contradictory—they illustrate that no matter how the human brain has come together over time as a collection of different mental modules, all plugged together, that collection is honed by evolution to serve its owner best, no matter what the environment. A hundred thousand years ago gatherers remembered where the edible food sources were; today barmaids remember who's having the Cabernet. On the other hand, knowing that the beer in the glass is horizontal has no apparent survival value now, and probably never had any. It does not come "spontaneously to mind." We can be trained to figure it out—a tribute to the brain's potential—but it doesn't have the mysterious, unconscious, intuitive quality exemplified by remembering what one hundred and fifty people are drinking at New Year's. That's survival.

Curiosities of Life

The Invention of Thievery

It might be the way science is taught in school ("this is the answer you should get"); it might be the media presentation of science as a series of brilliant new discoveries and insights (ever read the headline "Harvard scientists announce they have absolutely no promising leads . . ."?); it might even be scientists themselves. Whatever the reason, most of us non-scientists see science as an enterprise geared to producing answers, the more definitive the better. Each new experiment is expected to bring a solid new step in understanding. Big questions such as "Where did the human species come from?" or "How did the universe begin?" might seem like immovable objects in this view, but science is even chipping away at those.

There is some truth to this picture, but it's painted with much too broad a brush. Important among the details overlooked are those scientific questions that not only have not yet been answered, but realistically may never be. In some cases, especially those that represent one-time occurrences in the past, scientists can circle the mystery, getting ever closer with each new idea or experiment, but still never be sure they have the solution. This chapter is about such a phenomenon.

In 1921 in the English town of Swaythling a remarkable natural moment was witnessed. A tit—the English cousin of our chickadee—was caught in the act of drinking milk from a bottle left on a doorstep. Somehow the bird had pried open the cap and was enjoying the creamy upper layer. In the decades that followed, this novel habit spread, with reports coming from the length and breadth of Great Britain.

For more than fifty years scientists have been trying to explain both the invention and spread of this behaviour. It came as a surprise, because it represented a radical departure from the ordinary feeding behaviour of these birds (they mostly eat insects) and required such inventive dexterity. Because it is impossible to duplicate the England of the early twentieth century, scientists today have only two sources of information: records of eyewitness accounts of the phenomenon (of which there are more than four hundred in all) and modern birds with which to experiment. While these fall very short of being able to go back to Swaythling, researchers have been able to combine them in ingenious ways to offer plausible explanations for why tits suddenly realized that milk bottles were a good source of food. Inevitably though, such explanations lack the kind of certainty that many expect, and that leaves the door open for at least one fantastic and bizarre suggestion.

But before the theories come the data. From this single original observation in the early 1920s, milk-bottle opening spread inexorably across England into Scotland and Ireland. Through the 1930s the habit remained clustered in three areas, one near London, and two in the Midlands. By 1941 tits in Ireland had been seen doing it; by the end of the war there had been reports from all over England, several in Ireland and a couple in Scotland, and by the end of the 1940s tits everywhere in Great Britain were opening milk bottles.

In some places the spread could be traced in detail. James Fisher and R.A. Hinde, the scientists who collected the eyewitness accounts, noted that the behaviour was recorded four times in the northeastern suburbs of Belfast during 1937 and 1938. The habit then spread along the city's main roads and through small gardens; by 1947 there had been forty-seven incidents reported in Belfast. But that kind of spread—gradual and covering only a limited area—was only part of the story. If milk-bottle opening simply represented this local spread on a national level, there might not be a problem explaining it. Indeed, some animal behaviour experts have argued that it looks as through one creative bird came up with the idea of dipping into milk bottles, and the habit rippled outward from him or her. But there is a catch.

As Hinde and Fisher pointed out as far as back as the 1940s, tits are relatively sedentary birds, travelling no more than about twenty-five kilometres in their lives (and the scientists termed even that distance "exceptional"), and most of that during the first year. For the milk-bottle habit to have spread across Great Britain at the pace it did, there would have had to have been a crusade by hundreds of apprentices to that original clever bird. But if that had happened—a big if—why was the spread so uneven? The three areas in which bottle-opening was firmly established by the 1930s were widely separated. Why didn't the idea diffuse smoothly and steadily outward, with recruits popping up everywhere as it went? Surely there weren't counties in England where the tits tried milk bottles, didn't like them and abandoned the habit?

Despite the difficulties in explaining the spread of this habit as having sprung from a single original inventor (or a handful at most), some support this theory, probably because anything else seems even more unlikely. For example, the most reasonable alternative is that tits all over Great Britain figured out independently—at roughly the same time—that milk bottles were good food sources. This accounts for gaps, but if this were the case, why didn't the habit pop up in many locations at the same time, rather than spreading gradually north and west from the south of England?

There is more information on this phenomenon, although none of it addresses the central mystery directly. Hinde and Fisher noted in their records that the birds used a variety of techniques to get the cream out of the bottle. Most of the time a bird would attack the milk bottle shortly after it was left on a porch (although there were reports of the birds pursuing the milk truck down the street and ripping the tops of bottles waiting to be delivered). If the cap were metal foil, the bird would usually peck a hole in it, then tear the metal off in strips. Cardboard caps could be pulled up by the tab, removed intact, or thinned out layer by layer until a hole could be pecked in the middle. Some tits would even remove a cap and peck away at the cream stuck to the underside—piles of milk-bottle caps were found under trees or behind hedges.

Milk-bottle-opening was a blend of innovation and routine, because while the techniques of opening varied from bird to bird, habit was also important. There were several reports of tits which opened only milk caps of a certain colour. Exploiting this novel food resource wasn't without risk: the occasional tit was found drowned in a milk bottle, having reached down that one milli-metre too far.

How could these birds have had this stroke of genius? Hinde and Fisher, the original investigators, suggested that breaking into a milk bottle required some of the same skills tits have used for millennia in their natural environment, such as stripping bark off a twig to get at insects hiding underneath or hammering open a nut. In their eyes, the innovative step was the recognition of the milk bottle as a food source, not the technique for opening it. And that was about as far as they could go. Oddly enough the next steps were taken by Canadian scientists who, despite being geographically and temporally removed from the original event, nonetheless were able to design experiments that shed some light on the mystery.

In one case, David Sherry of the University of Western Ontario and Bennett Galef of McMaster University tested the innovative capacity of the chickadee, the North American cousin of the tit. In their first experiment, they presented captive black-capped chickadees with restaurant cream containers sealed with foil. Four of sixteen birds given this opportunity figured out how to open the container and get at the cream. Those are admittedly small numbers, but one out of four is a pretty good success rate for birds which, as far as anyone knows, had never seen cream containers before. But Sherry and Galef went on from there to test the learning ability of the other twelve birds, those who weren't creative enough to open the containers themselves. They might not have been inventive, but were they clever?

Four of the twelve were placed in cages divided in half by wire mesh. They were on one side; a "tutor" bird (one of the original inventive four) was on the other, easily visible. Both birds were given sealed containers of cream. Three of these four birds followed the example of their tutor and managed to open the cream containers, although only one performed successfully on every one of five trials. The fourth bird never did get it.

Some chickadees learned the trick after they were placed in a cage with an already opened container with a peanut and a sunflower seed inside. They succeeded in associating the cream container with food and were able to take the next step and open the real thing in later trials. This raised an important point: if a bird can make the connection between an open cream container and food, then it doesn't need to see another bird opening it. In

England, large numbers of opened milk bottles, on their own, could have triggered the behaviour in naïve birds.

This in turn raised questions about the role of the tutors. What was it exactly about the presence of a bird who already knew how to open cream containers that inspired birds who had failed to exploit the opportunity the first time around? Sherry and Galef set up a second set of experiments in which their test subjects again shared a cage with a tutor, separated from them by wire mesh. In some cases, the tutor was furnished with a container: he opened it and drank the cream as expected. In other cases, the tutor was there but had no cream to open. The surprising result was that the mere presence of a tutor on the other side of the cage —doing absolutely nothing—was as effective a teaching tool as having that tutor demonstrate the trick. Seven chickadees learned when they could see the tutor opening the container; six learned just by having the tutor there. All thirteen of the successful birds had failed to figure out what to do in a previous test, so it is unlikely they had just experienced a eureka moment. Unfortunately there was no way of knowing just what influence the tutors had in these experiments. Were they looking hungrily at the cream container on the other side of the mesh, or was their presence enough to galvanize the other bird into action?

Taken together, these two experiments suggested that the spread of milk-bottle-opening throughout Great Britain needn't have required role models to be copied. The mere presence of other birds—regardless of what they were doing—or already opened milk bottles could have played an important role as well. But as intriguing as these results were, and even though they suggested strongly that tits could come up with the idea independently, they still didn't address the central mystery: Was there one original inventor, or did this great idea somehow occur to many birds at roughly the same time?

The Invention of Thievery

In 1995 Louis Lefebvre, a biologist at McGill University, added a unique approach to the question. He asked whether it was possible to discriminate, armed only with the statistics of the spread of the milk-bottle innovation, between a single original inventor or roughly simultaneous inventions at a variety of locations. To evaluate the possibility that there had indeed been one or two incredibly brilliant tits, he compared the spread of milk-bottle-opening throughout England with an event from the much more remote past: the spread of agriculture through Europe from its origins in the Middle East. Although there is still some controversy over exactly how it happened, the archaeological evidence suggests that agriculture began to spread westward at a rate of about one kilometre a year, beginning about 6000 BC. There is no evidence that agriculture was independently invented anywhere in Europe during that time.

The agricultural data supports a steady and inexorable movement from east to west. When Louis Lefebvre applied the same statistical analysis to the spread of milk-bottle-opening, it just didn't fit. Even when he relaxed the criteria a little by adding a second possible origin to Swaythling—a sighting in county Durham in 1926—the data still didn't fit. Milk-bottle-openings just popped up too quickly in widely separated sites for there to have been any kind of bird-to-bird contact. On the other hand the data did fit with the hypothesis that there were multiple sites of innovation, followed by local spread from each. So the Belfast data mentioned above would represent one tit figuring out the milk-bottle gambit, followed soon after by dozens of birds picking it up, directly or indirectly from that original entrepreneur.

Recently Lefebvre has added to the picture by showing that innovations in feeding occur often in birds, and those birds with bigger forebrains are more likely to be creative. The innovations are sometimes startling: there are records of English sparrows learning to operate automatic door-openers, birds learning to pick insects off car radiators and even one account of magpies learning to dig up potatoes following the example of a single bird. While tits aren't as brainy as crows, ravens or birds of prey and therefore not as likely to innovate as those geniuses of the bird world, they are still brainy enough to have invented the milk-bottle trick. (Incidentally there are reasons to suspect that the habit might now be in decline: a Princeton biologist has suggested recently that these birds are probably unable to digest the lactose in milk and so were after the cream only, which is about 90 per cent fat with only traces of lactose. He further predicted that as the trend continues towards homogenized and skim milk (they now account for more than 50 per cent of the market), tits will be forced to look elsewhere for their fat intake. The same biologist, Carlos Martinez del Rio, also suspects that although Fisher and Hinde didn't specify which colours tits favoured in the past, that in future they will pick those that indicate the bottle contains non-homogenized milk.)

This brings us to an unsatisfying position, where all the evidence that is likely to be collected is already in, analyses more inventive than might even have been expected have been performed, but the definitive solution still eludes us. When that happens, as I suggested earlier, unorthodox explanations rear their heads. And the representative of the unorthodox in this case is the "theory of formative causation," an invention of a British scientist named Rupert Sheldrake.

Formative causation is the strange notion that there are invisible and non-material fields surrounding all of us, and indeed all living things on earth, called "morphic fields." These are reservoirs of accumulated experience that every member of every species draws on during its life. Sheldrake doesn't claim to know anything about the physical nature of these morphic fields, but he uses them to explain phenomena that otherwise are, at least to him, unexplainable. Morphic fields are particularly handy for explaining mysterious innovations.

According to the theory, any innovative act by an individual member of a species changes the morphic field of that entire species, even if their population is spread around the world, a process called morphic resonance. He defines morphic resonance not as an exchange of energy, but as an exchange of information. The effect of changing the morphic field is to make it easier for another member of the species to come up with the same innovation. That second or third or fourth individual need not have had any contact with the originator—they are led to the act by the alterations in the fields around them.

An example of the influence of morphic fields would be the nursery rhymes we are all familiar with regardless of our first language. Sheldrake would argue that because millions of humans have learned them before, they are much easier to learn now than they were when first told. He would explain the gradual rise in IQ test results over the past century in the same way (although there are other orthodox but convincing explanations). A more pertinent example is a case cited by Sheldrake of learning in experimental rats. In a thirty-year series of experiments begun by American scientist William McDougall in the 1920s, rats were trained to escape a water maze by choosing one of two gangways. Each led to dry land but one, which was illuminated, gave the rats an electric shock, the other didn't. The rats had to learn to escape using the unlit ramp. McDougall tested generation after generation of rats and found that each learned the maze sooner than the one before it. The first generation of rats needed an average of 165 trials to learn; by the thirtieth generation, they needed only twenty trials. Other investigators picked up the series of experiments and obtained essentially the same results. Rupert Sheldrake claims that these findings have never been adequately explained by orthodox science, but would be, as he says, "just what would be expected on the basis of morphic resonance."

I'm sure you can sense a connection about to be made. In his book *The Presence of the Past*, Rupert Sheldrake argues that the way milk-bottle-opening spread throughout Great Britain makes it a prime candidate for formative causation generated by morphic resonance. And indeed it is! After all, the habit spread in ways that aren't yet fully explained (even though experiments have suggested a combination of inventiveness and learning) and the idea that morphic fields surrounding tits made it slightly easier for each new generation to figure out milk bottles ties the whole story together. Sheldrake has even suggested that these fields extended out over the English Channel, making it possible for birds in Holland to start opening milk bottles immediately after World War II.

There is, however, a problem. Even though Rupert Sheldrake has been arguing for the presence of morphic fields since the early 1980s, and even though it is possible to design experiments that might demonstrate their existence, none of those experiments seems to have been carried out in a way that might convince a skeptic—a reasonable skeptic at that. Sheldrake himself admits that supporting evidence for his theory might never be found. I agree with him there. My brief encounters with Rupert Sheldrake (during an experiment of his that involved viewers of the Discovery Channel in Canada) left me with the impression that he is at his best when devising theories, not following them up.

Even his claims that he has the best explanation for William McDougall's mysterious rat experiments ignore some details that morphic resonance cannot explain. It is true that each successive generation learned the maze more easily, but—and Sheldrake doesn't mention this—after about twenty-eight generations, the rats started to make more and more mistakes. This shouldn't happen if the rats' "field" was being strengthened. Also a different set of experiments, with an easier task, revealed no evidence of improvement with time.

Rupert Sheldrake can argue that his ideas are too challenging to the orthodoxy and so will never be taken seriously, but if experiments have been done, where are the results for all to see? Tests of Sheldrake's formative causation were carried out in the mid 1980s; he referred to them merely as a "promising start"; in his 1995 book *Seven Experiments That Could Change the World,* morphic fields are still no better than a notion.

So we are left with two alternatives: solid data from good experiments that nonetheless leave us short of an explanation, or a theory that provides an explanation but lacks any experimental support. Which do we choose?

When I reached this point in researching this story, there seemed to be one important piece of the story missing: when did home-delivered milk bottles first make their appearance in England? If tits took to milk bottles as a natural extension of their normal feeding habits, then milk-bottle-opening should have begun soon after milk bottles appeared on doorsteps. If it didn't—if for instance there had been a period of decades during which the birds ignored the bottles—a more unusual explanation might be necessary.

From what I could find, it seems as if the tits were on the case pretty well right from the beginning. Anita Bourne of the National Dairy Council in Britain informed me that the first milk bottles, pioneered in the 1890s by Express Dairies, had wired stoppers which would have been extremely difficult to break into, but that when door-to-door milk delivery became more widespread in the 1920s, cardboard caps were introduced. Arthur Hill of the University of Guelph then dug out some corroborative evidence: in 1892, J.J. Joubert Limitée in Montreal had been the first dairy in the British Empire to deliver milk with cardboard caps.

The Barmaid's Brain

So cardboard caps first appeared in Canada and only arrived in England around 1920; the first report of a tit stealing milk was in 1921 and the story unfolded rapidly from there. I think these facts put to rest much of the remaining uncertainty about the origins of the habit. I would be willing to bet that the spread of the habit across England would correlate with the spread of cardboard caps, but I can't say that categorically without exact dates for the introduction of home milk delivery across the country. But I think that's why the habit popped up in remote sites, and while I can't be absolutely certain, it's a better explanation than formative causation.

In any case, I don't think we need a definitive explanation today. Very often phenomena that have no good explanation are that way because they are not well documented, or because fraud is suspected. In this case neither is true, but the explanations at hand still aren't fully convincing. So I declare a hung jury on this one. No answer—yet. And I'm content to wait for one.

The Plant That Rolls

———————

———————

The inner circle of animals that we find most attractive are those that have (at least in our minds) human attributes: they are cuddly and cute (koalas), fierce (grizzlies), immensely strong (elephants), cunning (foxes) or playful (otters). Birds—and their watchers—form their own unique group. Sharks, cobras and black widow spiders and animals like them fascinate us too because they are threatening. Most of the rest of the animal kingdom live their lives unnoticed by humans. Too bad, because among them are creatures that are at the same time alien and familiar, who live in a world utterly unlike ours, yet may provide clues to our own past— our *distant* past.

One such organism is Volvox (the "fierce roller"), and it is not even an animal. It is a plant, an alga. Volvox is a spherical being about as big as a grain of sand. In this size range, the visual distinction between "plants" and "animals" amounts to intracellular anatomical details, not gross differences in appearance. The clearest sign that Volvoxes are plants is that they are green.

The Barmaid's Brain

Anytime you splash your way through shallow muddy puddles in the Canadian spring, you're trespassing on prime Volvox habitat. (Although by leaving watery footprints, you may have created more.) They appear relatively briefly in such short-lived shallow bodies of water, and have for centuries captivated those observers who care to find them. Anton van Leeuwenhoek (pronounced "Lay-ven-hook"), famous in the seventeenth century for his pioneering microscopical discoveries, reported seeing Volvoxes for the first time in 1700, when he was nearly seventy:

> When I brought these little bodies before the microscope, I saw that they were not simply round, but that their outermost membrane was everywhere beset with many little projecting particles . . . all orderly arranged and at equal distances from one another; so that upon so small a body there did stand a full two thousand of the said projecting particles.
>
> This was for me a pleasant sight, because the little bodies aforesaid, how oft soever I looked upon them, never lay still; and because too their progression was brought about by a rolling motion . . .

Leeuwenhoek went on for pages about Volvox, describing weeks of observations of them. But none of the detail in that writing obscures the fascination he experienced upon seeing them for the first time, a fascination still felt by the handful of late-twentieth-century scientists who study Volvox.

Leeuwenhoek's observations through his handmade microscopes were perfectly accurate. The Volvox sphere does roll through the water, and close examination would reveal some two thousand "particles" studding the surface. Those particles are the cells that make up the organism. It is a colonial multicellular being, and therein lies its fascination. While a single cell lives an

The Plant That Rolls

*A plant the size of a grain of
sand that roams free through
the water; a colony of some
two thousand individuals
who, without communicating
with each other, somehow
co-ordinate their every action.*

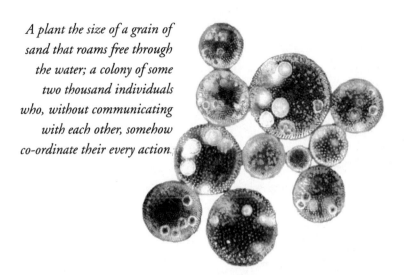

independent existence, the cells in Volvox are part of a co-operative. Some scientists think that understanding Volvox might reveal the secrets of that most dramatic of life's transitions, one that each of us made: one cell dividing many times to create a single organism comprised of its descendants.

As Leeuwenhoek wrote, individual Volvoxes are about the size of grains of sand. At first glance each sphere is reminiscent of a Buckminster Fuller geodesic dome, the surface punctuated by the individual cells Leeuwenhoek called "particles." In the most-studied species, there are some two thousand of these cells immobilized in a gelatinous extracellular matrix. Each of these cells has within it all the machinery necessary for life, including a chloroplast to generate energy from photosynthesis, an "eyespot" (not a real eye but a spot of light-sensitive pigment), and two whip-like flagella at one end. These flagella provide the first curiosity.

All the scientific evidence points to the fact that the cells in the Volvox community are nearly identical to—and certainly descended from—the free-living cells of an alga called Chlamydomonas. Same eyespot, same chloroplast, even many of the same genes. But a close look at those flagella gives a hint of the complexity of the shift from a unicellular to a multicellular existence, a shift that occurred at some point in the remote past. In Chlamydomonas the flagella are mounted at the front of the cell and beat in opposite directions, breast-stroking the cell through the water. The cell also rotates as it goes, allowing a constant sampling of the ambient light so that the cell can move towards or away from light depending on the circumstances. Now imagine these same cells—two thousand of them—arrayed on the surface of a sphere, flagella projecting out like tiny outboard motors. Each cell is free no longer—as the colony moves so does the cell. The flagella must beat in a way that moves, not each individual cell, but the *colony* in the right direction. To this end, the positioning and orientation of the flagella within each cell have been altered—they are turned around so that they beat in the same direction and in unison. The impact of this re-engineering can be clearly seen if a single Volvox cell is separated from its companions: it spins uselessly in the water, its flagella incapable of readapting to a solitary existence.

But fiddling with the flagella in each cell isn't enough. Remember that there are some two thousand of these cells on the surface of a sphere that moves through the water north pole first. That means that some cells are at the bow and some at the stern (the south pole), while the rest occupy various other positions in between. But all have to co-ordinate to move the sphere forward and rotate it at the same time.

I used to think that all the cells in the colony would have to be in communication with each other to ensure the co-ordination of those beating flagella. If the colony has to move up through the water column towards the light, shouldn't those cells everywhere on the sphere actively co-ordinate their flagella so that the colony moves in the right direction? I assumed that had to be the case, but apparently the colony swims smoothly and efficiently with absolutely no communication among its cells. In most species there isn't even any physical connection between them. Yet the green sphere moves purposefully and steadily through the water, rotating gently as it goes.

It is a trick of geometry. The individual cells are not haphazardly placed in position on the sphere; they are arranged in a well-engineered symmetry. Imagine the oarsmen of an ancient Roman galley given the task of propelling a spherical submarine. The oars protrude in all directions from the interior of the sphere. Each individual is concerned only with his own rowing, but the location of his seat and its orientation ensures that his efforts, together with his crewmates, will move the sphere through the water, rotating as it goes. This is the secret to Volvox. Indeed, mutants with cells that lack the correct orientation thrash their flagella wildly, jiggle madly and go nowhere.

It's a strange thought: a collective that operates in perfect harmony without any awareness (even biochemical) of the other members. Contrast this to the Borg collective of Star Trek. Trekkies would recognize the similarity instantly, but obviously it is superficial only. The enhanced humanoids of the Borg are all part of a communications network; each individual is under constant guidance and supervision. The Volvox collective is more subtle—it gives the *impression* of communication and supervision but requires neither. Its collective behaviour is determined by its structure: a place for every cell and every cell in its place.

While Volvox hit on a geometrical solution for the problem of getting places, the rest of its life is a little more complicated. The single-celled algae that are its predecessors live a routine life of swimming, growing, then splitting into two and repeating the process. In Volvox those cells have become two completely different kinds, neither of which is quite like the ancestor. One, the soma, makes up the vast majority of the colony's cells: their role is to beat their flagella and keep the sphere on the move. In specializing for this role, the cells of the soma abandon growth, and most important, their ability to reproduce. In fact, as the colony nears the end of its life, these cells commit suicide.

This isn't a trivial act: setting aside the individual capacity for reproduction seems to run counter to the evolutionary goal of ushering your genes (or copies of them) into the next generation. But it isn't suicidal altruism: the soma move the colony into areas of brighter sunlight for more efficient photosynthesis, or into deeper waters to absorb minerals. These activities improve the survival chances of the colony's offspring, all of which share their genes with these hard-working but infertile colleagues.

The other kind of cell is a germ cell; unlike the soma, germ cells are specialized completely for reproduction. They actually shed their flagella, leaving the responsibility for movement entirely to the soma. Volvox scientists are especially interested in this division of labour. How exactly were these two very different kinds of cells created from one?

Some light can be shed on that question by following the sequence of events from one generation of this organism to the next. This is tricky to describe because there is no one starting point—life as Volvox is continuous. The handful of germ cells (maybe no more than sixteen out of two thousand) in the mature sphere divide repeatedly and the products of those divisions, the daughter cells, form hollow clusters bulging inward towards the centre of the sphere. These are embryonic Volvoxes. Even though these embryos are still attached to the inside wall of the mother colony, the cells in them are already looking to the future, differentiating into germ and soma. There is one major hurdle to overcome: the geometry of the cell divisions has meant that the flagella on those daughters face inward, towards the centre of the embryo.

If left that way the Volvox sphere would end up lying motionless in the mud with its internally directed flagella madly stirring up the water inside it.

The young Volvox performs a neat developmental trick to cope with the situation. There is a small hole at the end where it remains attached to the mother sphere, and it simply turns itself inside-out through that hole, like a sweater being pulled right-side-out.

It's worth thinking about that for a minute. Remember it's assumed that the cells in the Volvox colony were once single-celled independent algae. It's hard to imagine that there could have been anything in their genetic repertoire that would have prepared them for this colonial contortion. Yet it is now simply part of their routine.

After turning inside-out, the daughter colony detaches itself from the wall of the mother and swims around inside, joined there by several other daughters, many of which already contain embryonic Volvoxes of the next generation. At some as-yet-undiscovered signal the parent colony breaks apart, its soma cells commit suicide, and the daughters are freed.

This is by no means a complicated life cycle, but it has enough within it to intrigue scientists, in particular the split between germ and soma. Beginning with one cell in the parent colony, an embryo develops in which appear, after several cell divisions, two distinct kinds of cell. One is immortal through reproduction; the other sterile and destined to die in the service of the colony. There must be a set of genes that determines this difference, genes that are switched on or off in the appropriate cells at the right time. Discovering what these genes are and how they work might reveal something of the mechanisms that drive the development of all embryos including human. How does an organism begin life as one omnipotent (scientists use the word "totipotent") cell and end with, in our case, more than two hundred different kinds of cells, most of which are concerned with maintenance, not reproduction.

David Kirk of Washington University in St. Louis is one of those rare researchers who has focused on Volvox for most of his academic life. He and his colleagues have made some progress in identifying the genes that play a role in differentiating soma from germ cells. One such gene regulates cell division. Early on in the developing sphere cell divisions are symmetrical: both daughter cells are equal in size. At a crucial point, however, cells located towards one end of the embryo begin to divide asymmetrically, giving rise to one huge daughter cell and one diminutive one. The huge daughter cells will become reproductive cells; their smaller sisters will be soma. There seems to be a critical threshold to that size difference. Kirk has found that any cell in the species of Volvox he studies that is more than eight-thousandths of a millimetre in diameter will become a reproductive cell; any cell that is less automatically becomes soma.

That is mysterious biology: soma and germ cells specialize by shutting down certain biochemical pathways and starting up others. Their identity is their chemistry, their repertoire of molecular capabilities. In our bodies the same is true: brain cells make neurotransmitters, stomach cells digestive enzymes. But in Volvox, chemistry is somehow tied to the sheer cellular volume. And it seems that size is a determining factor, not just a byproduct of spcialization. If unusually large Volvox cells are produced artificially, either by microsurgery or by the premature interruption of cell division, those cells become reproductive cells—even if they live in an area of the embryonic colony that only produces soma.

There are likely many genes involved in the generation of a Volvox colony. It's not exactly that these genes have appeared out of evolutionary thin air; some of the Volvox's relatives begin life as soma-like cells then transform themselves into reproductive cells partway through their lives, but it still requires some creative gene swapping, shuffling and mutating to come up with an organism that incorporates two completely different kinds of cell within a single colony. It's exactly those genetic processes that scientists like David Kirk are pursuing.

And why bother evolving a colonial pond-water alga anyway? What are the shortcomings of being a Chlamydomonas, the single-celled predecessor of Volvox? One apparent advantage is safety. There are dangerous predators in fresh-water ponds, some of them filter-feeders, whose beating cilia (smaller versions of flagella) sweep currents of water through their bodies, capturing the unfortunate organisms caught in the current. Chlamydomonas are small enough to fall prey to filter-feeders, but Volvox aren't. (This may account for the fact that Chlamydomonas is often found living in damp soil, where filter-feeders cannot exist.)

Also the co-ordinated beat of thousands of flagella give the sphere a higher potential top speed than its single-celled relatives, a feature that can be crucial for reaching places where they can live. Volvoxes in an African lake have been seen descending tens of metres down the water column in the evening, only to rise again the following morning, sometimes at an incredible (for something this size) speed of five metres per hour. It's suspected that such migration is necessary because the organisms need both sunlight and phosphorus. In the sunny surface waters there's not enough phosphorus, and it's too dark to live in the phosphorus-rich depths.

Being super-motile may be an advantage, but Volvox seems to have paid a high price for it. After all, the colony must direct food and energy to the construction of close to two thousand soma cells, which, while they work very hard propelling the sphere, contribute not one single offspring. In those same waters there are species in which every cell reproduces. The suspicion is that the soma cells are not just waving their flagella, but are at the same time absorbing essential nutrients that can then be distributed to the reproductive cells. All of this must add up to reproductive success, however, and it appears in some circumstances that Volvox colonies can produce more offspring intact than would have been produced had the mass of the colony been converted completely to isolated reproductive cells.

It would be a mistake to think of Volvoxes as primitive, as a kind of early and not very sophisticated attempt to develop a multicelled organism. In fact they are likely not more than thirty million years old or so. In other words, Volvox appeared for the first time long after the dinosaurs had disappeared, and at about the same time as primates first appeared on the earth. They are a *recent* evolutionary experiment. It's still too early to tell whether they have hit on many of the same mechanisms for converting single cells to colonies as were developed hundreds of millions of years before them or whether Volvox developed their own idiosyncratic way. I'd bet on the latter.

It's a cinch van Leeuwenhoek never imagined the evolutionary and developmental issues faced by the little green spheres he found in water from the pond. But even in 1700 these organisms had more going for them than mere charm. Leeuwenhoek made a point of remarking that the daughter colonies swimming inside their parent (and the granddaughters inside them) were evidence that life came from life: it was not the product of a mysterious spontaneous generation, as some of his contemporaries were arguing and continued to argue for another hundred and fifty years. The continuity and evolution of life in a tiny green swimming sphere—so much contained in so little.

Consumed by Learning

One of the most sensational, peculiar and ultimately frustrating and puzzling episodes in twentieth-century science played out in psychology labs in the 1950s and 1960s. It began with a claim that tiny worms could learn; controversial enough because these were invertebrates, primitive animals without backbones deemed too simple to be capable of learning. But that was only the beginning. The same researchers then made the jaw-dropping announcement that they had been able to transfer memories from one animal to another, simply by allowing an untrained one to eat the chopped-up remains of another which had already been trained.

It was sensational science while it lasted; it ended, not with a dramatic experiment demonstrating that memories couldn't be transferred (the way things are supposed to happen in science) but with the scientific establishment turning away to pursue other research. Not only did the majority doubt that worms could learn by eating other worms; most couldn't be bothered proving those experiments wrong. They may have been turned off as much by the personality of the scientist at the centre of the controversy as by the science; it is almost certain that the final explanation will never be clear-cut.

The story began in the 1950s with the work of an American psychologist James McConnell and his colleague Robert Thompson. They had apparently done with flatworms what Pavlov had done with dogs, that is, established a conditioned response. Pavlov's dogs salivated at the sound of a bell; McConnell's flatworms scrunched up at the flash of a light. The dogs were conditioned to perform that unusual act by having food presented together with the sound of the bell. The flatworms first experienced the light together with an electric shock, and eventually the light alone was sufficient.

I've made it sound pretty straightforward, but a flatworm is a long way from a dog (or at least from most dogs). McConnell worked with planaria, the commonest free-living flatworm (its relatives include common parasites like tapeworms). Planaria are found in freshwater streams everywhere. They are less than an inch long (fifteen to twenty millimetres or so), are indeed flat, and unlike earthworms, have an identifiable head and tail. The head is triangular with two obvious eyes. They secrete a slime trail that facilitates gliding around on stones or sand or even on the underside of the surface film of still water.

They aren't much internally, just a mouth that opens on the underside of the body midway from head to tail, a simple digestive tract which runs both directions from the mouth, and most important in this case, a simple nervous system. It consists of a concentration of nerve cells in the head (that might just qualify as a brain) and two separate nerve cords that run from the brain all the way to the tail end of the animal. The planarian is unique in that it is the simplest animal to have a "brain."

In their experiments McConnell and Thompson set up a plastic trough, not unlike a section of eavestrough, with water in the bottom and an electrode at each end. Two 100-watt light bulbs were mounted directly above the centre of the trough. A planarian was taken from the aquarium, placed in the trough and given ten minutes to acclimate. Usually at the end of that time the animal would be gliding smoothly back and forth along the trough.

Then the unsuspecting animal was subjected to the conditioning routine: the light was switched on for three seconds, accompanied during the last second by the electric shock. McConnell and Thompson found that their planaria gradually responded more and more to the light by both turning their heads and contracting their bodies; neither response became more common if the worms were exposed either just to the light or the shock alone.

Those experiments were published in 1955, and while they were important in demonstrating that planaria could be conditioned in the classical sense, they weren't spectacular. McConnell took care of that oversight a few years later when he demonstrated that the memory created by the conditioning had some extremely peculiar properties. To do this, he took advantage of the worm's ability to regenerate. If a planarian is sliced in half, the head end will regenerate a tail and the tail will grow a head, with two intact worms as the result. Depending on its size, a worm can be cut into dozens of pieces that will regenerate new worms.

So what happens, asked McConnell, if a worm is taught to react to the light, and then that worm is cut in half? Do the regenerated worms retain the memory of the light and the shock? Well, they do. Using the same experimental set-up, McConnell trained a set of worms (when they were "fully trained," they would respond to the light at least 23 times out of 25 trials), then chopped them in half and allowed them four weeks to regenerate. At the same time, other worms were cut in half without being conditioned first, and a third group was conditioned, then simply allowed to rest for four weeks.

There is nothing complicated about these flatworms, but they can learn simple tricks. Thirty years ago scientists were convinced those tricks could be fed to others.

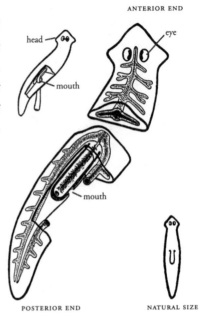

The worms regenerated from the head end reached the status of "fully trained" much sooner than they had when they were originally trained, meaning that the memory had been preserved. That might not have been so astonishing, given that the head end contained the planarian brain. But explaining the other part of the experiment wasn't so straightforward: the worms regenerated from the tails also appeared to have retained the memory, in some cases better than the heads.

What could a memory be made of that would allow the tail end of a trained worm to remember? It had to be some sort of chemical, but obviously it wasn't resident in the brain alone. But the idea that a memory might be contained within a molecule of some kind that could migrate freely around the planarian nervous system was a bit too much for some of McConnell's fellow psychologists to swallow. It was at this point, in 1959, that doubt and opposition began to grow, although much of it remained unpublished and even unspoken. It wasn't to stay that way long, however, because McConnell's list of spectacular experiments wasn't over.

He and some of his co-workers wondered if it would be possible to transfer whatever it was that encoded planarial memory from a trained to an untrained animal. They first tried grafting a trained head onto an untrained body, but such grafts didn't take very often. Then they tried injecting ground-up animals, but that didn't work either. Finally they hit on the perfect scheme: under the right conditions, planaria turn cannibalistic. And given the primitive nature of their gut, it might be possible that memory stuff could enter an untrained cannibalistic planarian relatively intact.

So they embarked on the most controversial experiment yet: training worms, chopping them up and feeding them to hungry untrained worms. (In anticipation of the reaction to these experiments, they instituted a number of control groups, emphasizing "blind" procedures to ensure that the people testing the animals didn't know what kind of animal they were testing.) The amazing result was that cannibal worms performed significantly better on test runs after they had eaten trained worms, although this instant learning decreased over the following few days. McConnell admitted in one of the first publications of this data that "we are most definitely out on a limb of some kind," although it's unlikely he knew just how far.

There were two kinds of opposition to McConnell's work, formal and informal, and it is pretty clear that the informal was just as damaging to his work and his reputation. For one thing, many scientists simply didn't believe McConnell's results, because they didn't think they could be true. A perfect example was a meeting McConnell had in the late 1950s with invertebrate expert Libby Hyman of the American Museum of Natural History. When told that regenerated heads and tails retained their conditioning,

Dr. Hyman said, "I'm very sorry but I just can't believe that." When McConnell remarked that the planaria sometimes retained their memory for weeks, she replied, "No, that just can't be. I could believe that a planarian might remember something for five minutes or so. But weeks or months? No, that just can't be." This story was related by McConnell himself, but there were undoubtedly many others who didn't believe him but never admitted it to his face.

The idea that something can't be true because it's impossible to envision *how* it could be true surfaces again and again, particularly in medical science. Those in the medical profession who don't believe in, or are antagonistic towards, complementary medical treatments like chiropractic, acupuncture or therapeutic touch are opposed in part because they can't imagine how they would work: how can stimulating an acupuncture point in the foot affect the eye? Where is the so-called aura said to be manipulated in therapeutic touch?

In the case of the planaria, it wasn't so much the surprise that planaria could learn. That could be accepted, and was, but no one expected memories to come in nicely packaged molecular form, especially a form that could be moved from one animal to another. There were already strong opinions that memories involved some form of rewiring in the brain, which wasn't something you could feed your neighbour. In the light of these beliefs, memory transfer was somewhere between audacious and unbelievable.

However, holding a personal opinion is one thing, but in the court of science evidence is primary, and that means replication. Controversial experiments should be repeated in an open-minded way by other scientists in other labs. Either the results hold or they don't—at least that's the way it's supposed to go. But it's rarely the case that a scientist will repeat another's work to the letter, and as soon as the experimental details differ, uncertainty and controversy walk in. That uncertainty characterized the scientific follow-up to McConnell's work is a deep understatement.

It's difficult to be completely accurate about how many people were able to replicate these experiments and how many weren't, due at least partly to the tendency for other groups to fiddle with the original experiments. However, as you read the scientific literature through the 1960s, you can't avoid the uneasy feeling that whether the evidence against McConnell's work is clear-cut or not, the tide is turning against him.

The first published discrepancies appeared in 1964. A California group tried to replicate the cannibalism experiments and found that worms which had eaten trained worms did better all right, but so did worms that dined on worms exposed only to light or even worms just handled by the experimenters. There seemed to be no sensible explanation of these results, but at the very least they called into question the reliability of McConnell's results.

In the same year a different team reported they had failed to replicate McConnell's results, and the particularly damaging aspect of these experiments was not the results, or even the conclusion they drew (a relatively mild "learning was not yet proven"), but who did them. One of the two-man team was Melvin Calvin, a Nobel laureate in chemistry. Once a laureate contradicts experimental results publicly, those experiments are in trouble.

A second refutation was published in 1966. Although it was specifically aimed at the extension of McConnell's experiments to rats, and although this letter published in the journal *Science* examined only one possible way of transferring memory, it dealt a heavy blow to all memory transfer research. It was signed by twenty-three researchers affiliated with seven different institutions, all of whom had failed to find any evidence of the chemical transfer of memories. Even when I was writing this chapter, and asking researchers to remember what it was that brought down

McConnell's work, several remembered—sometimes vaguely—this letter in *Science*. (Some of these were the same people who told me they couldn't remember what happened because they never believed it in the first place—replication or not). Yet even the *Science* letter concludes with a plea for continued work in this area: "We feel it would be unfortunate if these negative findings were to be taken as a signal for abandoning the pursuit of a result of enormous potential significance . . . Failure to reproduce results is not, after all, unusual in the early phase of research when all relevant variables are as yet unspecified."

Despite these noble words, the letter writers could hardly have been surprised that their letter was seen as a critical blow against memory transfer research. Certainly their remark about "all relevant variables" was appropriate to the planaria research. At one point there was an estimate that some seventy variables could have influenced the experimental results, including the size of the worms, the colour of the light, the nature of the electric shock, the time of day, the phase of the moon, you name it. Planaria weren't easy to work with, and McConnell and his co-workers found that practice with the worms, and learning to recognize when they were ready to learn, was an important part of the procedure. But that meant certain people were better at the experiments than others, and situations like that always raise suspicions. And any time McConnell or one of his colleagues tweaked an experiment a little to get better results, they were accused of searching for positive results. Scientists are opinionated, and such opinions can be a lens through which failure to replicate results can either be seen as sloppy or fraudulent work or as an intriguing phenomenon. The bottom line is that by the very early 1970s the planaria work had been abandoned, even though some researchers who lumped all the experiments together and analyzed them came up with more positive results than negative. That's science.

Consumed by Learning

One other factor in this story was James McConnell himself. He was unusual, iconoclastic and very public, none of which stood him in good stead. In 1959 he founded a new journal called *The Worm Runner's Digest*. You have to see it to believe it. Alongside scientific experiments there are cartoons, little pieces of personal philosophy, jokes, satire—it looks like anything but a science journal, yet some of the planarian research appeared there and only there.

In 1967 McConnell went halfway towards dignifying *The Worm Runner's Digest* by creating *The Journal of Biological Psychology* and packaging the two together, back to back. The *Journal* had the serious research, the *Digest* the fun. But even then it must have been hard to take it completely seriously.

McConnell was also a tireless public communicator, a role that scientific colleagues often resent (David Suzuki and the late Carl Sagan experienced plenty of that). He might also have gone too far for his own good. Mark Rilling at Michigan State University has pointed out that at the 1959 meeting of the American Psychological Association, McConnell talked to his peers about a memory molecule, but then to the press about a "memory pill." Shortly afterward *Newsweek* proclaimed that "It may be that in the schools of the future students will facilitate the ability to retain information with chemical injections." Then in 1964 *The Saturday Evening Post*, in a feature on McConnell, reported that his research might "someday enable us to learn the piano by taking a pill." Obviously none of these articles would have endeared McConnell to other psychologists, even though in his work environment he disclaimed the idea of a memory pill.

In a final, bizarre twist to the story, McConnell was the target of an assassination attempt by the Unabomber on November 15, 1985. McConnell wasn't gravely injured by the letter bomb, but his hearing was permanently impaired. It's impossible to know exactly why he was targeted, but Mark Rilling suggests that it might have been McConnell's bold public endorsements of behaviour modification that attracted the Unabomber's attention. Rilling cites, for instance, an article in *Psychology Today* called "Criminals Can Be Brainwashed Now," remarking that this contains exactly the message the Unabomber found distasteful: using technology to change society.

James McConnell died in 1990. To my knowledge no one trains planaria to do memory transfer experiments any more. Some of the people who did left the field not so much because they didn't believe in it, but because their peers were suspicious of it. Today's psychologists believe memories to be permanent changes in the synapses, the connecting points between nerve cells. There are indeed molecules that must be synthesized in association with those changes, but no one thinks they can be fed to others and create "learning." There really is no unambiguous evidence that any molecules like that ever existed.

But when you look back over the entire saga, it seems memory transfer was not so much disproven as abandoned. In the words of sociologists Harry Collins and Trevor Pinch, "We no longer believe in memory transfer but this is because we tired of it, because more interesting problems came along, and because the principal investigators lost their credibility. Memory transfer was never quite disproved; it just ceased to occupy the scientific imagination."

Why Do Moths Fly to Lights?

There are many scientists who specialize in the study of insect flight. Some put moths in wind tunnels and add a little smoke to make visible the whorls of air flitting off the ends of the moth's wings, giving them lift. Several of them specialize in bondage, gluing insects to various pieces of lab equipment, then watching them try to fly. I, on the other hand, collect good stories about insect flight, and here are my favourite three: two are unanswered questions and I'll save them for later; the first actually contains a rare example of science humour, by which I mean a scientist using science to be funny.

Irving Langmuir was a Nobel-winning chemist with little patience for sloppy thinking (he coined the term "pathological science" for the "science of things that aren't so," warning that excessive accuracy, ad hoc excuses and barely detectable effects are all signs of such pathology). At various times in the 1930s, he was troubled by claims in the media that deer flies could fly seven or eight hundred miles per hour, but the source of the claim eluded him until the New Year's Day 1938 edition of the *Illustrated London News*, which quoted Dr. Charles Townsend, writing in the *Journal of the New York Entomological Society* back in 1927:

> On 12,000 foot summits in New Mexico I have seen pass me at an incredible velocity what were certainly the males of *Cephenomyia*. I could barely distinguish that something had passed— only a brownish blur in the air of about the right size for these flies and without a sense of form. As closely as I can estimate, their speed must have approximated 400 yards per second.

It was this last claim that intrigued Langmuir. Four hundred yards per second is 818 miles per hour, comfortably faster than the speed of sound at sea level, an incredible velocity for any living thing. There was no indication that there had been a slip of a decimal place in Townsend's report, and Langmuir was moved to dig deeper. He published his analysis of Dr. Townsend's report in *Science* magazine in March 1938, beginning with an equation.

The equation came from the 1929 *Encyclopaedia Britannica* section on Ballistics, and it showed you could calculate the drag on a flying object from data like its speed, its shape and the density of the air. When Langmuir applied this equation to the deer fly's 818 miles per hour, he found that the drag on the fly would be a force equal to a hundred grams. That may not sound like much, but remember it's being applied to a relatively tiny area: the head of a fly. The result is that the fly would have to endure forces of about eight pounds per square inch, enough, Langmuir figured, to crush it.

Langmuir then tried to calculate the fly's energy consumption, beginning with the estimate that the fly would burn half a horsepower, or 370 watts, just maintaining its unbelievable pace. Unfortunately that would necessitate the fly consuming one and a half times its own weight . . . every second. He further showed that if a deer fly hit you at 818 mph—as deer flies on the wing often do, and with some force—it would penetrate deeply into your flesh.

Finally Langmuir designed some simple experiments to explain the flight of the deer fly. He wondered what Townsend had meant by a barely distinguishable "blur in the air." So he did what any practical scientist would do: he fashioned artificial flies from fly-sized chunks of solder with silk threads tied around their middles, then whirled them around in the air. Once he was sure he could measure the velocity of the "fly" as it orbited him, Langmuir watched it at various speeds. He reported that by the time it was moving 13 mph it was already a blur; by 26 mph it was "barely visible as a moving object" and at 43 mph it was just a streak in the air. Finally, by 64 mph the "fly" was invisible. Langmuir concluded that: "The description given by Dr. Townsend of the appearance of the flies seems to correspond best with a speed in the neighborhood of 25 mph."

I have never come across a response from Dr. Townsend, but then again, what would he say? He was a competent entomologist who had related the deer fly story in an article in which he argued that studying insects would be valuable in designing newer and faster aircraft. In fact he had been criticized by others before Langmuir caught up to him, but never with such wicked thoroughness. Langmuir was an unbelievably productive and inventive scientist, an utterly dogged man who never stopped trying to understand the natural world and who would wrestle with a scientific problem until he solved it. Apparently this approach to life agreed with him: he once said, "I have never spent a really unhappy day in my life." I can only assume that destroying Townsend's claims made that particular day even happier.

Whirling artificial flies around, matching their velocity to their appearance, reminded me of physicist Richard Feynman's famous gesture of plunking a chunk of O ring material from a space shuttle into a glass of water to demonstrate how it became brittle as it got colder. Ultimately these booster seals were blamed for the explosion of the Challenger in 1986. Both demonstrations are vivid and simple. Both are good science. And Langmuir's is funny, albeit with a cruel tinge; he would have been perfectly at home on talk shows in the 1990s.

Langmuir, after enduring the rumours about the deer fly's speed for years, was finally able to identify Townsend as the source and put the story to rest. On the other hand, I (admittedly with none of his scientific acumen or tenacity) have been unable to pin down the source of another piece of bug mythology, and (again unlike Langmuir) I'm unhappy about it. Some time early in the twentieth century someone—somewhere—made a claim that you occasionally will hear even today: "Scientists have proven the bumblebee can't fly." It's supposed to be a funny stab at the (hypothetical) nerdy nonsocial lab-bound researcher with no clue about what goes on in the real world. I have a photocopy of what appears to be the most complete version of this. It looks as though it was originally a single page in a book or magazine:

THE BUMBLEBEE CANNOT FLY

according to the theory of aerodynamics and as maybe readily demonstrated through laboratory tests and wind tunnel experiments, the Bumblebee is unable to fly. This is because the size, weight and shape of his body, in relation to the total wingspread, makes flying impossible.

BUT . . . the Bumblebee, being ignorant of these profound scientific truths, goes ahead and flies anyway—and manages to make a little honey every day.

Now I know a scientist didn't write that: not because it's witty but because a scientist of either sex would know that most of the bumblebees that fly are females, not males, and that neither sex makes honey. But it is clearly an attack on pointy-headedness.

A scientist may not have written it, but there is some truth in what was said. Bumblebee flight, at least in the middle of the century, was not explainable by the standard equations used in aerodynamics. But that doesn't mean that the scientists thought they were right and the bumblebees were wrong—they knew they just had more work to do. My photocopied essay turned the story on its head, making the experts look stupid, and aroused my curiosity: where did this weird story get its start?

I'll tell you right now: I still haven't found out the source of this quote, but here is how close I have come: in March of 1989 a scientist named John McMasters published an article in *American Scientist* magazine dealing with several topics, one of which was the origin of this claim. He recounted an incident that was supposed to have taken place in the early 1930s in Germany. There was a biologist and an aerodynamicist at a dinner party (where was the rabbi to complete the joke?) and the biologist asked the aerodynamics experts if he could explain how bees' wings worked. The aerodynamicist, after scribbling a few numbers on a napkin (or something) could only conclude that the standard aerodynamic theory indicated that bumblebees should not be able to fly. (There is another version which has the aerodynamicist overheard by a member of the media who immediately made the outrageous claim public—of course.)

In McMasters's article he refused to attach a name to this aerodynamicist, other than to claim he was a Swiss professor famous for research into supersonic gas dynamics. But he has backed off that claim more recently, saying that he has changed his mind about the Swiss professor, quoting instead some lines from a 1934

French book called *Le vol des insectes* as representing another version of the origin. He has added that there were probably several origins and his was "as serviceable as any."

Unfortunately *Le vol des insectes* said nothing at all about bumblebees. The author, M. Magnan, does write: "I applied the laws of air resistance to insects and I and Mr. Sainte-Lague arrived at the conclusion that their flight was impossible." However, further along he says, "One shouldn't be surprised that the results of the calculations don't square with reality." Somehow that just doesn't match the story of the Swiss professor at the dinner party: it's too vague (there's not a bumblebee in sight) and more important, it just doesn't have the sparkle such a story should have. I'm hopeful the real version might come to light one day.

In the meantime there is one loose end of this story that *has* been tied up. Scientists now know what the embarrassed aerodynamics expert at the dinner party didn't—how bumblebees fly. The standard aerodynamics used by the mystery expert made unrealistic assumptions that simply don't apply to insect wings: it treated them as if they were flat, rigid wings moving steadily at a constant speed. But they are not: they bend, twist, slice into the air and most important, generate little puffs and swirls of air that couldn't have been predicted by standard science. These mini-whirlwinds roll along the underside of the wings and generate enough additional lift to keep even a heavy-bodied insect like a bumblebee in the air. So the problem wasn't the short-sightedness or arrogance of scientists—they knew their approximations weren't adequate. They just needed time and experimentation.

And finally: wouldn't you think if scientists can explore the first few microseconds of the universe or direct a robot spacecraft around on the surface of Mars, they could give you the simple explanation for something that every Canadian sees just about every summer night? You'd think so, but I'm not convinced that the experts know why moths fly into lights. Many entomologists will tell you there's no puzzle at all, that there's a solid explanation that was put forward decades ago, but there are also some puzzling observations that don't square with that explanation. But judge for yourself.

In the late 1930s, a German entomologist named Von Buddenbrock (who likely attended the same dinner party as the aerodynamicist) argued that moths, most of which are night-flying creatures, would likely use the moon as a navigational tool. An airborne moth could maintain a straight course (which makes more sense than flying aimlessly in circles) by adjusting his direction of flight so that the moon always remained in the same place in the sky. For instance, to fly east when the moon is high in the southern sky the insect would keep the image of the moon on his right (many of them are males looking for females).

This would be a simple but reasonably reliable system that would have evolved with moths over millions of years. It's not perfect: the moon does move across the sky during the night, so an insect would have to compensate for that movement if it were aloft for many hours; if not, its straight line would become a gradual curve. However, the advent of artificial outdoor lighting would have thrown a major wrench into this neat system of navigation. Buddenbrock showed that if a moth fixates on a porchlight instead of the moon, it will try to maintain its bearing relative to the light instead of the moon. The geometry of the situation predicts that the moth will begin to circle the light, but rather than settling into a stable orbit will instead spiral inward until it actually crashes into the light.

That is the standard explanation—it's in all the textbooks—and there is some evidence for it. Some species of moths apparently use the moon as a reference point for long-distance flight. Robin Baker and his colleagues in England demonstrated this with some elaborate experiments on a moth called the large yellow underwing (which has nearly three thousand relatives in North America). It might surprise you that determining whether moths follow the moon should require "elaborate" experiments, but these are night-flying moths—you can't see them after they've flown a few metres away. Somehow you have to be able to determine the direction they *would* fly but at the same time keep them in view.

So in an experiment reminiscent of Irving Langmuir whirling tethered chunks of solder around his head, Baker's team built a vertical stand with an arm extending out from the top (like a gallows); a moth with a thread glued on its back was suspended from the arm. The moth was free to flap its wings and head off in any direction; when it did so, its movements turned a spindle which triggered electrical switches, thus recording how it moved.

They chose the large yellow underwing because it showed signs of being a moth that flies great distances. On one occasion they used a light trap (a trap with a light bulb as the lure) to collect 791 of these moths in an hour, far more than could have been living in the neighbourhood. They must have been flying in from some-where. For these experiments Baker set the apparatus up outdoors in a place where it appeared (at least from those moths they could watch at dusk) that the insects tended to fly in straight lines, ap-parently unaffected by lights in the windows of buildings close by.

The results were dramatic. In the presence of a nearly full moon, the tethered moths flew, or tried to, in straight lines. How-ever, the moment a screen was erected to block the light of the moon, their flight became more-or-less random, "multi-direction probing flight" as Baker called it. Then after the moon had moved behind some trees, a 125-watt artificial light was intro-duced to the plot, placed about two metres away from the moths. They dutifully switched their flight direction to orient to the light, maintaining the same angular relation with it as they had with the moon.

These results seem to substantiate Buddenbrock's theory that moths mistake artificial lights for the moon and blunder into them. One of the curious twists here is that the brightness of the artificial light didn't seem to be the most important factor. The reaction of the moths depended more on the elevation and size of the light. If the light was only 60 centimetres above the ground a moth had to be within about three metres before it would be attracted. But if that same light were raised to a height of nine metres, approaching the height of a three-storey building, moths even fifteen or seventeen metres away flew towards it. At that distance the image of the light is about the same size (in the moth's eye) as the image of the moon.

Baker's experiments are the closest thing to proof that moths mistake artificial lights for the moon and fly into them. Were there not anecdotes and experiments that don't fit this theory the case would be closed. But there are, and one of the best is a particularly elaborate example of the "insects in bondage" theme. In this experiment, H.S. Hsiao designed a moth flight pond to determine the insects' reaction to light. Yes, a *pond*. He glued (do you think they used special insect glue?) corn earworm moths to tiny styrofoam boats floating in a tray of water with a light suspended above it. The liquid in the tray (two layers of water with a layer of oil in between) was electrified, so that when the moth boat moved, its position and direction were recorded on a graph. This design allowed the moth to propel its boat around unencumbered by wires.

The results were not quite what you might have expected. Not surprisingly, with the light off, the little boats wandered aimlessly around the pond. With the light on, the moths dragged their little craft towards the light, as expected, but they did not exhibit the spiral path predicted by the theory. Instead most of them actually headed not for the light itself, but to places on either side of the light, with a few individuals heading directly into the light and banging into it. Hsiao concluded that moths do not fly to the light, but rather to a place just next to it. His experiments tally with the data from light traps in the outdoors, where it seems that the greatest number of moths collected is some distance from the light, perhaps a half a metre or so, depending on the species caught. I, too, have noticed that if I leave the outside light on at my cottage at night, the walls in the morning are covered with moths, many of them perched metres away from the light.

What is this all about? Hsiao suggested that his experiment could be best interpreted by assuming that the moths were initially attracted to the light, but then at the last minute were trying to escape it. Why wouldn't they then turn around and flee instead of heading for the slight shadow near the lamp? Hsiao could only suggest that it might have something to do with the way their eyes work—research on other animals like horseshoe crabs has shown that their eyes and brain together exaggerate slight light-dark differences, in this case turning the slight shadow into darkness. But there is no evidence that this mechanism works in moths' eyes, and this whole adventure in the moth pond leaves more questions than it answers. Why do moths suddenly switch from approaching the light to avoiding it? Are they bedazzled by the light's intensity when they get too close? On the other hand, this experiment mimics the findings from the light traps, where the majority of moths are not caught next to the light, where the standard theory predicts they should be.

While this has all been straightforward science so far (if you can call watching moths propel little styrofoam boats across an electrified pond straightforward), this field does include a touch of the even more bizarre. It begins in the nineteenth century with the great French entomologist Jean Henri Fabre. In the book *The Insect World of J. Henri Fabre* he describes an amazing night that he called the Great Peacock evening. The Great Peacock is the biggest moth in Europe, a gorgeous insect with a huge eye-marking of chestnut purple and black on each wing. One May morning in Fabre's house a female Great Peacock emerged from her chrysalis and he immediately placed her inside a wire-mesh bell-jar and then went about other business. That night around nine o'clock his son suddenly called out, "Come quick! Come and see these Moths. Big as birds. The room is full of them!"

One of the windows of the study where the female moth was caged was open and male Peacocks had invaded the house. Fabre estimated there were at least *forty* of them, all attracted by the prospect of mating with the newly arrived female. Fabre, a good scientist, spent the next week trying to understand how the males had realized she was there. He concluded that the attraction couldn't be visual—there were just too many trees and shrubs obscuring the window, especially in the dark of night. He likewise rejected sound and, after some hesitation, smell, and wondered if the female was emitting some kind of "luminous radiation" to which the males are sensitive. ("Does she, in her own manner, employ a kind of wireless telegraphy?") To that end he amputated the antennae of some of the males on the suspicion that these might be the "radiation" detectors, indeed most of those males did not return on the night of their antennal loss.

Entomologists today would interpret that little experiment differently. The female, once emerged, began to release a come-hither chemical called a pheromone, an airborne chemical (not radiation) that male moths use their antennae to detect. They are so attuned to it that a single molecule can bring a male moth from great distances. Amputating their antennae, as Fabre did, would render them unaware of the pheromone and unlikely to return to the house.

But the most fascinating moment of the Great Peacock evening was when Fabre brought a candle into the study to see what was happening:

With a soft flick-flack the great Moths fly around the bell-jar, alight, set off again, come back, fly up to the ceiling and down. They rush at the candle, putting it out with a stroke of their wings; the descend on our shoulders, clinging to our clothes, grazing our faces.

They rush at the candle, putting it out with a stroke of their wings.

What does this mean? These male moths were in the overwhelming presence of a high concentration of female pheromone and their mission in life is to mate. Why would they be distracted by the light of a candle? Even if candles were not part of their evolutionary history, there should be no reason why they should be attracted to one. In fact it happened again with a paraffin lamp enclosed by a white enamel shade. Nine male moths came seeking the female; two lit on her cage, the other seven perched on the lamp shade and remained there until morning. Fabre wrote: "The intoxication of light had made them forget the intoxication of love."

The moths' attraction to light, as oddly powerful and puzzling as it was, was a frustration for Fabre, who chafed at the difficulty of studying an animal that changed its behaviour the moment you turned a light towards it in order to see it behave. He gave up working with the Great Peacock. However, Fabre's speculations about a mysterious radiation passing from female to male, even though they are ignored by most entomologists today, have not died completely. In the mid-1960s a scientist named Philip Callahan put a radical—some would even say nonsensical—suggestion on the table. His idea might be strange, but it also makes the attraction of the moth to the flame much more exotic. He claimed that the attractant molecules released by the female, the so-called pheromones, actually do radiate energy to which male moths are sensitive. To be more accurate, they reradiate energy, absorbing low intensity ultraviolet from the night sky and giving it back in the form of energy in the infrared part of the spectrum. So while it is the same kind of energy as light, its frequency is too low to be visible—we perceive infrared radiation as heat, not light. The pheromone molecules radiate this energy simply as a by-product of their molecular structure.

Furthermore, according to Callahan, the males detected the female's pheromones, not by coming in contact with the actual molecules ("smelling" them), but by tuning in to the wavelength of their radiation. The antennae—the sensory extensions that Fabre had lopped off in the name of science—were antennae in the true Radio Shack sense. In fact, Callahan's theory started with the antennae. He predicted from their form, arrangement and even their electrical properties that they were receivers, and that the female's pheromone chemicals would produce the infrared radiation those antennae had evolved to detect.

Having put forward this theory, which represented a dramatic departure from anything before or since, Callahan produced data suggesting that moths are attracted to lights, especially candles, because the infrared radiation from the candle is indistinguishable, at least from the moth's point of view, from the radiation that is emitted by the female's sexual pheromone. To the male moth, according to Callahan, a candle flame looked the same as a female moth, although "looked" isn't quite the right word—maybe "felt" is better. The females of different species would produce their own pheromones emitting characteristic frequencies of infrared; the males of that species would have an arrangement of branches and spines on their antennae that would be tuned to those frequencies. If the poor male flies into the blizzard of light from a candle, he circles hopelessly, unable to pinpoint the source.

Callahan has even demonstrated how the spectrum of radiation from a candle, a Bunsen burner and a Coleman lantern differ, with prominent peaks in the right part of the infrared for the candle and the Bunsen burner, but none in the spectrum on the Coleman lantern. He asserts that moths are attracted to the candle and the burner, but not to the lantern, and then argues that since moths should be able to see the visible radiation from the lantern, but remain unattracted by it, then they must be homing in on something else.

Callahan first put forward this theory in the mid-1960s and it has gone . . . nowhere. Most entomologists will admit having heard of it, but at the same time dismiss it as irrelevant or unproven or just wacky. For his part Callahan argues that most entomologists don't know physics as he does and so either don't read his work, or don't understand it. While it is true that I have never found a definitive rebuttal to Callahan, chemist Martin Moskovits at the University of Toronto pointed out that while the theory isn't impossible, moths using this infrared radiation would face a number of difficulties, the most prominent of which is that the night sky is a cacophony of infrared. Males would have to be able to tune out that background and tune in to the relatively weak signal from the female's pheromone, a task Moskovits thinks is likely too demanding to be feasible. Also each molecule is emitting infrared, so the atmosphere around the female would appear to be (to coin a phrase) a thousand points of light.

As far as I can tell, Callahan's theory doesn't explain why moths might end up turning away from the light at the last moment, unless the radiation intensity close to the flames simply overloads the poor male's nervous system. Regardless, Callahan's ideas remain largely unaddressed in the world of insect experts and I fear his is what surely must be the worst fate for a scientist: no one is listening to him.

There are some other suggested explanations for the moth and the flame: the moth heads towards light because it represents open sky and escape from danger, or the heat from the candle or light-bulb heats up the near wing of the circling moth, causing it to beat faster and thereby circle ever closer, but there are no experiments to back up these minority views. And this is where the mystery stands today. I say mystery, because while there is evidence that moths mistake artificial lights for the moon, I'm still not convinced that their behaviour close to the light is explained best by that hypothesis. The Fabre story is a good example: why would male moths intent on mating ignore a female to fly into a candle? Surely that would be a choice that evolution would never favour, although I suppose it could be argued that the choice between a candle and a female never arose in the moth's evolution. That story does make Philip Callahan's ideas more intriguing.

There are entomologists who will tell you that we are only on page one when it comes to explaining why moths fly into candles, which suggests a lot more gluing is in order and further, that we should leave the last word, for now, to a poet, Thomas Carlyle (please note however: for "she," read "he"):

With awe she views the candle blazing;
A universe of fire it seems
To moth-savante with rapture gazing,
Or Fount whence Life and Motion streams.

What passions in her small heart whirling,
Hopes boundless, adoration, dread;
At length her tiny pinions twirling,
She darts, and—puff!—the moth is dead.

Homo Aquaticus

To maximize the vividness of this chapter, you should be standing naked in front of a full-length mirror when you read it. Reading it in the bathtub might be even better. If neither appeals to you, then at least cast your mind's eye over your body, and consider the following list of features.

First and most obvious, its nakedness. Our closest relatives, the great apes, are covered with hair. Of all the species of mammals, only a handful (among them manatees, whales, dolphins, elephants, hippos, rhinos and pigs) are hairless. The interesting thing about the animals in that list is that several of them are aquatic, and some that aren't today might have had aquatic ancestors.

The hairs you have retained do not just stick out in random directions, nor do they run parallel down the body; instead they are arranged diagonally, pointing in towards the midline of the body, a pattern that would encourage the easy flow of water over your body if you were swimming.

Homo Aquaticus

The sweat glands scattered all over your body cool you when you're overheated, although they are by no means perfectly suited to that task: excessive sweating removes too much salt and water from the body, so much so that you need to consume expensive athletic beverages to restore the salt balance. In the days before such beverages, wouldn't it have been advantageous for our ancestors to live somewhere where both salt and water were in abundance?

If this point-by-point inspection of your body leads you to weep emotional tears, that too is a unique human trait, something that would also have been useful (for removing excess salt) had we once lived in a high-salt environment.

You can only infer its existence from a glance in the mirror, but there is a virtually continuous layer of fat under your skin, a layer of fat that is seen most obviously in aquatic mammals, like manatees, whales and walruses. That fat may even alter your body shape, making it more streamlined in the water.

Your nose sticks out prominently from the rest of your face, nostrils down, ideal for preventing water from getting in it when diving.

Imagine you are about to dive—the first thing you would do is hold your breath. That ability—to decide to hold your breath—is something no other terrestrial animal is capable of doing. Why would we alone have that skill, so useful when swimming?

The very fact that you are standing on two legs means you would be able to breathe in chest-deep water.

And now you can get dressed. In 1960, a British scientist, Sir Alister Hardy, was impressed enough by this suite of anatomical and physiological features to propose a radical alternative evolutionary history for human beings. He argued that these apparent traces of a watery past made the standard evolutionary picture of an ape-like creature (our ancestor) that moved from the African forests onto the savannah and stood up in the process highly unlikely. Instead, Hardy proposed that the ape-like ancestor moved out of the forest all right, but wasted no time moving into the ocean, lived there long enough to develop several adaptations for marine life, then returned to land some millions of years later.

Hardy first presented his ideas to a meeting of the British Sub-Aqua Club in Brighton in early March 1960. He titled his talk, "Aquatic Man: Past, Present and Future," but the emphasis was clearly on the past. He argued that our primitive hominid ancestors had been forced by competition to exploit a new ecological niche: feeding on seashores and in shallow waters. In an article in *The New Scientist* a couple of weeks later (written to clarify what Hardy thought had been misrepresentations of his speech), he wrote:

> I imagine him wading, at first perhaps still crouching, almost on all fours, groping about in the water, digging for shellfish, but becoming gradually more adept at swimming. Then, in time, I see him becoming more and more of an aquatic animal . . .

Hardy died in 1985, but his ideas have been kept alive—and extended—by others, notably the British science writer Elaine Morgan in her books *The Aquatic Ape, The Scars of Evolution* and most recently *The Aquatic Ape Hypothesis.* The aquatic ape theory is not a "respectable" scientific theory; it doesn't appear in anthropology texts, is rarely referred to in scholarly journals and most experts in the field—paleoanthropologists—spend zero time thinking about it. Even the Internet newsgroup devoted to it (sci.anthropology.paleo) petered out early in 1997 (and not a minute too soon, I'd say, after spending a few hours reading the mostly useless—and nasty—material in it).

I began researching the Aquatic Ape theory hoping it would contain the same kind of imaginative twists and turns that other off-the-beaten-track ideas provide, but it was exactly the reverse: the more I searched, the more dead ends and negativity I encountered. Even those few scientists who have published their thoughts on the theory in scholarly publications like the *Journal of Human Evolution* dismiss the aquatic ape theory as being just this side of crackpot. There is one exception: Daniel Dennett, the tough-minded philosopher famous (or notorious) among academics for his books *Consciousness Explained* and *Darwin's Dangerous Idea,* had this to say in the latter about the aquatic ape theory:

> When I have found myself in the company of distinguished biologists, evolutionary theorists, paleoanthropologists . . . I have often asked them just to tell me, please, exactly why Elaine Morgan must be wrong . . . I haven't yet had a reply worth mentioning, aside from those who admit, with a twinkle in their eyes, that they have often wondered the same thing.

The few experts I've forwarded this question to either politely refuse to comment or, in one case, complained that the question had been poorly phrased. So phrase it differently; Dennett's question is still worth asking. There are still big gaps in the scientific account of human evolution, and such gaps invite speculation and unofficial scenarios. The question that always must be asked is, do the alternative scenarios actually help fill the gaps?

Scientists spend a lot of time pointing out the similarities between us and our closest relatives, the chimpanzees. There are two species of chimp, the common chimp that Jane Goodall has studied for decades and the lesser known bonobo, or pygmy chimp. Both are genetically very similar to humans (the common chimp closer), so much so that some scientists think of humans as just another chimp (see Jared Diamond's *The Third Chimpanzee*).

But let's face it: there is world of difference between a human and a chimp. The most obvious is mental, notwithstanding the linguistic achievements of chimpanzees like Kanzi, the chimp trained by Sue Savage-Rumbaugh who apparently understands complicated English sentences (there are even structures in the chimp brain that hint at some sort of organization for language) or the reasoning exhibited by chimps who are smart enough to pile up boxes to reach bananas suspended from the ceiling. They're smart, but they're not *Homo sapiens* smart. And the difference between us and the chimps is more than just mental: physically and developmentally we're completely different animals. And yet there's that genetic similarity—the genes of the two species are more than 98 per cent identical.

That not only suggests that we are, at least genetically, a third chimpanzee, but that at one time in the not-too-distant past we and the chimps shared an ancestor. That ancestor is crucial to understanding how we evolved. The English evolutionary scientist Richard Dawkins has illustrated that relationship between us, chimps and that ancestor with an unforgettable metaphor.

He asks you to imagine setting up a human chain. He places his in Africa, but it could work anywhere. Imagine you're the first link in the chain, standing just at the western limits of the city of Montreal, facing north. Your left hand holds the right hand of your mother, her left hand is holding the hand of her mother, your grandmother and so on. The chain crosses the Ontario border and continues down Highway 401 towards Toronto. But amazingly, before it even gets to the Toronto city limits, it has reached the common ancestor we share with the chimpanzees. (Had you been standing on the outskirts of Winnipeg, the chain would similarly have stopped just short of Regina). Now if that ultimate ancestor (joined to you by an unbroken chain of half-a-million generations) then extended her left hand to grasp the hand of another daughter, the chain would begin to loop back on itself, this time moving east back along the 401 through time ultimately to reach another of her descendants, standing right beside you: the modern chimpanzee.

Dawkins created the idea of the continuous chain, each individual an immediate family member with those adjacent, to make the point that we and the chimps are indeed blood relatives. And while the chain illustrates that the accumulated differences between the two species have been exceedingly gradual, it doesn't minimize the size of those differences. It's exactly those differences that make the chain idea so astonishing. That common ancestor, the individual at the loop of the chain, is suspected to have lived somewhere around five to six million years ago in Africa. Events that we might never come to understand conspired to split that group of ancestral animals into two: one evolving into today's chimpanzees, endangered forest-dwellers, the other into humans, who live everywhere and are the agents, not victims, of endangerment.

That six-million-year span is fascinating. Who, or what, were the animals ancestral to us, what did they look like and what were they capable of? How did the story unfold to the present? Attempts to answer these questions have filled several years' worth of *National Geographic*s, and while a lot of progress has been made, the truth is that our understanding of that six million years' worth of evolution is porous to say the least.

Fossil skeletons are by far the most important evidence—without them we would have absolutely no idea of what went on. Fossils do double duty: they give us clues to what our ancestors looked (and acted) like, and they constrain the wilder speculations.

Sometimes a good fossil can force even the experts right back to the drawing table. When I first started reading about human evolution twenty-five years ago, it was generally believed that the hallmarks of humanness—upright walking, a big brain and dexterous hands—had evolved in a sort of circular fashion. The growing brain demanded that the hands be freed from locomotion so that they could fashion and use tools at the brain's bidding. Once that process began, an even bigger brain would design even more complicated tools, requiring even more manual skill, and the race to become a modern human was underway. Of course at a somewhat later point language was in there too, so that big-brained hominids could tell other big-brained hominids about their neat tools: the evolutionary roots of Black & Decker.

But then Don Johanson and his team discovered the famous skeleton of "Lucy," a more than three-million-year-old creature who, while already apparently walking upright, hadn't yet bothered with the big-brain part of the story, and certainly wasn't making any tools. So the idea of upright walking being forced into existence by other human attributes had to be thrown out. Clearly two-leggedness had come first. The discovery of Lucy spawned new theories about the origins of two-legged walking, and even now, more than twenty years after Lucy, there is no clear winner.

The problem is that Lucys only come along every few decades. There just aren't enough fossils to make the path of human evolution clear. Sure, there's Lucy and some even earlier relatives from Ethiopia; there are some fantastic footprints in volcanic ash from Tanzania and one or two nearly complete skeletons, plenty of jawbones, teeth galore. But there are very few hominid fossils older than three million years, and none at all older than four and a half million years. That leaves a million-and-a-half-year blank following the hypothetical chimp/human ancestor. To put that into perspective, a million and a half years ago the dominant hominid on earth was *Homo erectus*, a completely different and smaller-brained species than us.

The rarity of hominid fossils means that only about one in every ten thousand or so of those individuals in the hypothetical human chain lining Highway 401 has left behind any kind of fossil that we have found. No wonder the fossil of one diminutive upright-walking female upset the applecart. And no wonder we can speculate about an aquatic ape ancestor.

Two-legged or "bipedal" walking is only one of the ways we differ from the chimpanzees, gorillas and orangutans, our close relatives. Besides the obvious additions like language and really big brains, there are those features revealed or implied by the mirror: nakedness, projecting noses, a layer of fat. Is it justified to speculate that those are leftovers from a time when we lived in the water? The aquatic ape hypothesis makes many such claims, but a select few raise the most interesting questions.

One is upright walking. The fossils can tell us when our ancestors switched from four legs to two, but not why. The "why" must somehow be inferred from the locale of the fossils, the climate of the time and the other fauna sharing that locale. Even the famous Lucy, while proving that bipedalism preceded the dramatic growth of the human brain, displays some ambiguities in her skeleton. Her pelvis, hip, legs and feet suggest she could walk upright pretty much like you and me, yet the foot bones are still slightly curved, suggesting she and her kin might still have been spending time in the trees. But why were they becoming two-legged at all? Were they being forced down from the trees by a changing climate that was creating more open grassland at the expense of the forest? And even if they were, why resort to walking on two legs when four were available?

That last question is the most important. Sir Alister Hardy argued that living in neck-deep water would have encouraged such behaviour, if not forced it upon us, but most of the science that's been done to unravel the history of bipedalism has concentrated on the more traditional conjecture that forest-dwelling ancestors made the transition to two legs as they ventured out onto the plains. Some studies show that walking on two legs consumes much less energy than walking on four, but such comparisons are tricky: if you compare a chimp on two legs to itself on four, the two modes are about the same.

If you compare a human on two legs to a chimp on four, the human is much more energy efficient. But is that a sensible comparison? We have been adapting our muscles and skeleton to two-legged locomotion for more than three million years, but bipedalism had to have been an advantage to the first creatures who used it, however crude their two-legged locomotion was. No animal takes an evolutionary step so that its descendants will benefit.

There actually was another ancient ape that walked on two legs. Its name is *Oreopithecus bambolii*, and it lived on an island in what is now Tuscany as much as eight million years ago, even earlier than the chimp-human common ancestor. The skeleton of this one-metre-tall creature bears signs of upright posture—an S-shaped spine and long thigh bones—but its feet suggest it was more of a shuffler than a strider. So at least for this creature efficiency and speed weren't the most important factors, and this raises the unpleasant prospect (unpleasant for those who seek simplicity) that there can be more than one reason for standing up. Maybe bipedal walking appeared several times in evolutionary history—we don't know yet.

If the energy/efficiency differences between two and four legs weren't the crucial factor, what might have forced us upright? There have been several suggestions. Hominid social life might have been predicated on males roaming far and wide gathering food to bring back to the stay-at-home female to whom he was mated, and what better way of doing that than to free the hands for carrying? The non-sexist version of that idea is that the females brought food back to a central area—a camp—for processing.

Other theories include the idea that standing up allows you to see farther (although animals that do that today, like meerkats, are perfectly able to stand up, survey the scene, then return to walking on four legs), and an intriguing idea put forward by Pete Wheeler of Liverpool University that standing upright protected our ancestors from the blistering heat of the African sun because it exposes much less body surface to the sun's rays. But for that heat reduction scenario to make sense, there has to have been some reason for the early hominids to leave the forest cover in the first place.

These are all reasonable ideas, all contentious, and all without much evidence behind them. So what does the latest version of the aquatic ape idea suggest? In her updated book *The Aquatic Ape Hypothesis*, Elaine Morgan argues that we should acknowledge the importance of the fact that the skeletons of Lucy and her kin were found near the shores of a large lake which apparently was subject to periodic flooding. She suggests that walking on two legs would have made it possible to forage for food that had suddenly been inundated. In earlier writings she has argued that the problems we still experience caused by upright posture, including lower back problems, varicose veins and fainting, would have been alleviated by the buoyant effect of the water.

She points out that the proboscis monkey of Southeast Asia, which lives in mangrove swamps, is often seen walking on two legs to get from A to B at high tide. Morgan also cites observations of bonobos wading through streams and suggestions that this behaviour might account for their "long, strong legs" and relatively erect posture. You'll note (as so many of the theory's critics have) that this is a very long way from the full-blown version of the aquatic ape, which had our ancestors swimming, diving and generally behaving like hominid otters.

While it is true that most hominid fossils, and certainly the ones like Lucy exhibiting the final stages of transition to bipedalism, are found near water, that doesn't mean they were living in it. The problem is that bones are all we have and there is virtually nothing in bone that would indicate that a hominid was a water-dweller. Even Elaine Morgan points out that if you had only the skeleton of an otter and nothing else, you'd have no idea how much time that animal spends in the water. Webbed feet and water-repellent fur are not well-preserved in the fossil record. Unfortunately this is a weakness for the entire aquatic ape argument.

Our nakedness is another rare trait. We share it with the subterranean naked mole rat, elephants, hippos, rhinos, walruses, whales, dolphins, manatees, pigs and a couple of others. Elaine Morgan argues that with the exception of the mole rat, all these animals either are aquatic, like the manatee, whale, dolphin and walrus, or could possibly have had some acquaintance with water in the past. For instance, some evolutionists believe that the group to which elephants (and manatees) belong might have shared some sort of aquatic ancestor. It's pretty thin evidence, but the aquatic ape hypothesis needs it to tie nakedness to the aquatic life.

Alister Hardy felt that hairlessness made sense because a naked body moves through the water more easily; he even cited the fact that competitive swimmers who shave their bodies can swim (marginally) faster. (Hardy was so carried away with this idea that he argued that losing body hair would made the difference between being caught by a shark or escaping. A shark! And swimmers weren't even using steroids in those days.) But furry aquatic animals, like the polar bear, otter, beaver and a number of seals don't seem to be at any disadvantage in the water; why did some become naked and not others?

This nakedness stuff gets very confusing, because you can't really separate it from other human traits, like sweating and fat. Besides a bare skin, humans have a continuous layer of fat under the skin (so-called "subcutaneous" fat), and we sweat to cool ourselves off. So instead of protecting ourselves from the sun and/or warming ourselves with fur, we cool ourselves with sweat and keep warm with fat. But why did our ancestors go to those lengths when there are plenty of animals who evolved in comparable conditions who just kept their fur and got on with it?

The aquatic ape hypothesis has answers. When we were living

in the water, we adopted subcutaneous fat to prevent heat loss, which is much more rapid in water of a given temperature than in air. (Many fully aquatic animals today have a similar fat layer under the skin.) At the same time, we lost our hair, possibly to facilitate speed in the water. Upon returning to land, the problem of overheating under the hot African sun became much more pressing, and active sweat glands soon proliferated on the bodies of these humans-to-be. Of course all of these changes were presumably accomplished by the usual evolutionary mechanisms of natural selection: those individuals who had more sweat glands or more extensive fat had a slight advantage and so were the most prolific.

These are the most interesting scenarios sketched by the aquatic ape hypothesis, because they attempt to explain features that challenge even the standard accounts. The skeptics' counterarguments reflect that uncertainty. John Langdon of the University of Indianapolis, author of a recent scholarly treatment of this subject, simply says that nakedness is no better explained by the aquatic ape hypothesis. Langdon and other experts are on slightly stronger ground when they argue that our nearly continuous layer of body fat is merely an extension of the normal pattern in primates— where other animals have isolated patches or "depots" of fat, ours have run together. The fat in aquatic animals is different: it forms a continuous thick layer all over the body. They also point out that fat, while good for warming, is at the same time no barrier to cooling, because overheated blood vessels can bypass the layer of fat and carry heat to the skin where it can be lost. However, this does not explain why, if we did evolve exclusively on land, we completely abandoned the ancestral methods of heating and cooling for the ones we have.

But if our fat deposits are a legacy of an aquatic life, why are there differences between the sexes? Women deposit fat on their hips, thighs and breasts, a tendency that begins around adolescence. It's assumed that these fat deposits are sexual signals, indicating readiness and ability to reproduce—nothing to do with keeping warm. Elaine Morgan responds by taking these sex differences back into the water: perhaps females could have been the first to explore the water environment, while the conservative dominant males hoarded the rapidly diminishing food reserves on land. And let's not forget that fat increases buoyancy. I'd say the fat issue isn't resolved.

One final example: our ability to hold our breath at will is something that appears to be shared only by some aquatic animals, like seals and porpoises. Other primates are incapable of doing so. Isn't this a quintessentially aquatic adaptation? This is at first glance a difficult problem for aquatic ape skeptics, but there has been one explanation put forward. John Langdon quotes a study of breathing and locomotion that demonstrates that the musculature of quadrupeds makes it impossible for them to disentangle their breathing from their movement: as the legs move, their lungs contract and expand.

Put a rabbit on a treadmill, and it will breathe in a predictable manner: at high speeds, for instance, it takes one breath per stride. Upright walking, on the other hand, frees the upper body, including the diaphragm and chest muscles, from the legs and opens the door to voluntary control of breathing. The human on the treadmill may stride once every three or four breaths, or may make five strides for two breaths. Is this an adequate explanation? I'm not convinced. For one thing, voluntary control necessitates involvement of higher levels of the brain, areas concerned with decision making and thinking. There must have been evolutionary pressure for those areas to be recruited to this task even after two-leggedness had been achieved. What would that pressure have been?

Both sides in the dispute have pointed to speech as a unique human ability that depends on precise control of breathing, but full-blown speech seems to have arrived on the scene millions of years after the first bipeds walked the earth. That isn't to say that the eventual development of speech didn't depend on our breath-holding ability, but it doesn't explain why we acquired that ability in the first place. However, it would obviously have benefited an animal spending long periods of time underwater.

So where does the aquatic ape hypothesis stand? Critics attack it for making no useful predictions, for being so elastic that it can accommodate any piece of contradictory evidence simply by back-tracking (the swimming ape has become the wading ape) and add that there's absolutely nothing in any hominid fossils discovered in the last twenty years that would support it. They seem irritated that Elaine Morgan and other supporters of the theory shift their ground but refuse to abandon the theory. Of course those same critics neglect to mention that the establishment ideas about human evolution are a moving target too: no one today would use the same scenario for human evolution as was popular twenty-five years ago—Lucy ensured that.

So which is it? Are paleoanthropologists insecure with the gaps in their story and so take out their anxieties on the aquatic ape theory? Or is it really bad science—or not even science at all? There are two big problems with the aquatic ape scenario: it makes no predictions that can be tested and hominid fossils reveal no traces of an aquatic existence. That doesn't leave much: pointing out that we're naked like manatees and can hold our breath like seals doesn't mean we lived in the water.

On the other hand, the aquatic ape idea isn't totally useless: it raises some interesting questions about the standard view of human evolution. Rather than elevating it to the status of an alternative description of our evolution, maybe it should be viewed instead as a collection of speculations about selected peculiarities of our species. Some of these are farfetched, like the idea that downward-pointing nostrils evolved to keep water out when you dive from the fifty-metre board, or that acne is a leftover from a time when our aquatic male ancestors secreted oily sebum from glands on the head and neck to be able to slip through the water like a greased pig. But some, like breath-holding and subcutaneous fat, are not.

There is much about the generally accepted picture of human evolution that can't be proven, and perhaps the most important contribution the aquatic ape hypothesis can make is that it directs attention to the weaknesses in the tale, and forces scientists to admit those weaknesses and continue the search for better evidence. The proponents of the theory won't like this suggestion because it demotes the theory to a collection of scattered observations; scientists won't like it because they view the theory as being somewhere out beyond the fringe.

One final note: although the proponents of an aquatic past never cite this as hard evidence, from Alister Hardy on there has been, in the background, this thought: why are we humans so obsessed with the sea? Why do vacationers go to the beach? What about swimming pools and hot tubs? Are these not evidence of an innate bond to watery environments?

Although not intended to answer this question, studies of preferences for landscapes have revealed that regardless of cultural background, people shown photographs of a choice of landscapes preferred those with trees typical of the African savannah, a preference explained by evolutionary psychologists as harkening back to our (hypothesized) time spent there as hunter-gatherers those many years ago. However, the presence of water in such scenes makes them even more attractive. Water that reminds us of our past? Or water that we know to be cooling in summer, good to drink and a unique source of food?

Homo Aquaticus

The answer is not apparent in our brains. So far it has not been found in the bones either. But who knows? Answers to some of the questions raised by the aquatic ape theory may come from an unexpected direction. But be warned: they may not come for a long long time.

Science and History

Saint Joan

The 1990s was to have been the Decade of the Brain, as officially proclaimed by then-president George Bush. It has turned out to be a fantastic period of research (with still some time to go): new images of the working brain are published every week, chemical details underlying brain diseases promise better treatments, and even consciousness—the unique personal world that we all inhabit—is yielding some of its secrets.

It is tempting to apply this newly acquired knowledge to a handful of historical personages whose claim to fame or notoriety might be found in their grey matter. Bits of Albert Einstein's preserved brain are examined under the microscope to see if he had unusually large numbers of brain cells or more than the normal number of connections between them. Russian scientists have done the same for Lenin. Psychologists have tried to diagnose the erratic behaviour of Vincent van Gogh (was he manic-depressive?) and even Isaac Newton (mercury poisoning late in life?). But to me the most intriguing person—the one who presents the most challenging set of symptoms—is Joan of Arc.

Joan of Arc's career as a patriot, soldier and martyr lasted only two years but changed the course of French history and, centuries later, led to her canonization. An illiterate teenage peasant girl becoming a military strategist and leader would have been astonishing enough, but in all her adventures Joan claimed to be guided by the visions and voices of saints. She was certain that God was speaking to her directly; modern commentators with a secular bent would say she was hallucinating.

What makes this case even more fascinating is that there are plentiful records of what Joan did and said, both during her military campaigns and at the trial that led to her execution. The question is, what do they tell us about what was going on in her brain? One of the first medical experts to comment on Joan's life wrote in 1883: "As science shows us the relation of events previously unknown, we see more surely how things really occurred; what was perplexing becomes clear . . . " Nobody today, even with most of the Decade of the Brain behind us, should speak with that sort of assurance when it comes to Saint Joan.

The richness of her story is in the details of her life, but for that you will have to read one of the many biographies of Joan; I will hit only the highlights. She was born into a peasant family in the town of Domremy, east and a little south of Paris, in 1412. France at the time was divided: the north was held by the English, supported by the armies of the Duke of Burgundy. They had proclaimed the infant son of the English king Henry v and the French princess Catherine as King of France. Their forces controlled most of France north of the Loire river (approximately the upper third of modern France) and were besieging the town of Orleans, south of Paris; if they took Orleans, the south of France, still loyal to the dauphin, Charles VII, would likely fall. The dauphin, although heir to the French throne, was putting up no resistance to the English and remained isolated and uncrowned, in a castle in Chinon, to the southwest of Orleans.

Even the children of Domremy couldn't escape the political tension: the neighbouring town of Marcey was loyal to the Burgundians, while Domremy still held out for the dauphin. Gangs of boys from the two towns would occasionally meet and duke it out. But even though the villagers of Domremy had to flee before a Burgundian force once during Joan's childhood, her life was a relatively simple one of growing up with four siblings, tending to the family's sheep and cattle, learning to sew and attending church. It was this last activity that first set Joan apart from her friends. She was said to be extremely pious, confessing unusually often, so much so that she was teased for it, a remarkable fact in an age steeped in religion.

Her piety soon gave way to something more mysterious (or miraculous depending on your view). When she was twelve Joan experienced her first vision. At her trial, Joan remembered being in her father's garden at about noon when she heard what she described as a voice, seemingly coming from her right, in the direction of the church. The voice was accompanied by a bright light, also coming from the right. After she had heard the voice three times, Joan said she realized it was the voice of an angel.

A different account of her first voices is found in a letter written by a member of Charles vii's court, a Perceval de Boulainvilliers, who probably had access to an earlier statement Joan made to a commission appointed by the dauphin. The letter describes how Joan had been racing other girls across a field when she was stopped by a voice telling her to go home because her mother needed her. When Joan arrived at the house, her mother denied having sent for her and Joan, puzzled, started back to join her friends. A cloud passed in front of her and voice from the cloud told her that she must change her course of life and do marvellous deeds, for the King of Heaven had chosen her to aid the King of France. She was to wear men's dress, take up arms, be a captain in the war, and all would be ordered by her advice.

The two stories are significantly different, but this discrepancy hasn't troubled Joan's biographers. One account has her hearing the voice in her father's garden, the other on returning to her friends, at which point she could still have been in her father's garden. The story Joan herself told at her trial came two years after de Boulainvilliers heard the version he described—the differences in detail could have arisen for any number of reasons.

Are these fantastic stories to be believed at all? After all, Joan and Joan alone experienced the visions and voices, so she could simply be making them up. If she is not to be believed, the whole story becomes a grand deception. But her account of the visions and the role they played in her life (and through her how they changed the history of France) is just too consistent. It would be hard to find a good reason for Joan to make her visions up: they ultimately led to her being burned at the stake, a fate that clearly terrified her, and there isn't any evidence that she would have had to resort to claims of hearing voices to persuade soldiers, court officials or the dauphin to believe in her. There was much greater tolerance for the supernatural then than there is now. Joan's voices were accepted—in a way that is hard for us to understand—as part of the package, but she was judged on her determination, her actions and their success.

Her first vision was only the beginning. The voices stayed with Joan for the rest of her life, comforting her, giving her counsel, telling her what to do. And what she did was hard to believe, voices or no voices. Remember, she was an illiterate teenage girl in a war-torn country, yet she persuaded hardheaded military men to accompany her on a mission to the dauphin at Chinon. He in turn was impressed enough by Joan to allow her to recruit and ride with an army that set out to do what the dauphin and his followers could not: begin to take France back from the English. With Joan riding prominently at the front in full armour, her forces took one town after another; in a dramatic battle they broke the English siege of Orleans and soon after made good her vow to see Charles VII crowned King of France at Reims.

The dauphin was, however, not nearly as powerful a character as Joan and he vacillated when she urged him to move on to Paris. Her support weakened further when she was captured in an attempt to force the English to retreat from their siege of the town of Compiegne. Eventually Joan was sold to the English.

She was tried, not by a political court as an English prisoner of war, but by the church as a heretic, blasphemer, sorceress and idolater. After sixteen gruelling trial sessions, lasting more than a month in all, through which Joan had stood strong, she faltered at the prospect of being abandoned to the English and recanted, renounced her voices and promised to "live in unity with the church, nevermore departing therefrom." She had been counting on being allowed to live out her life peacefully in the company of women, but was shocked to discover that she was sent back to prison with the same male guards who had been tormenting her. She soon donned men's clothing again (something she had been ordered by the vice-inquisitor not to do) and listened once more to her voices.

When this fall from grace was discovered, Joan was condemned to burn at the stake. Her executioner, upon discovering that her heart had not burned, panicked, certain that he had killed a saint.

At her trial Joan had testified that she was visited regularly by three saints: Michael, Catherine and Margaret. She was able to describe them with some, but not complete and sometimes changing, detail: they spoke to her in French (when asked by her judges if St. Margaret spoke to her in English, Joan shot back, "Why should she speak English when she is not of the English party?"); St. Margaret and St. Catherine seemed to have appeared as heads with crowns but no discernible bodies or clothing and Michael's appearance was even vaguer—Joan wasn't sure what he looked like or wore or even if he had hair. But she claimed to have embraced Catherine and Margaret and that they had a sweet odour. It is unclear whether Joan could have described them in greater detail but simply refused, or whether they were indeed as vague as her testimony. There is some evidence that she was only able to produce detailed descriptions upon repeated interrogation.

The saints may have been indistinct in appearance but not in what they said. They visited Joan several times a week (sometimes several times a day), hundreds of times over seven years, and not only urged her to put her faith in God (advice Joan scarcely needed) but also mapped out her life for her. From the beginning her voices urged her to "go into France"; they told her to raise the siege of Orleans and have the dauphin crowned in Reims. They didn't just give orders, they made predictions: they warned her that she would be wounded in battle (she was) and Margaret and Catherine predicted (accurately) that she would be taken prisoner before the Feast of St. John. Early in her career, Joan shocked those still skeptical of her by announcing the French defeat at the Battle of Rouvray, days before the arrival of official confirmation of the battle. She was said to have had a revelation.

To complicate the picture, Joan sometimes acted independently or even in defiance of her voices. The standard she carried in battle (white with pictures of Jesus, angels and the self-conscious phrase, "on behalf of the King of Heaven") was apparently her design and hers alone. At no point does she credit her voices with it. In a peculiar incident after her capture, she leapt from a castle tower and fell an estimated twenty metres, escaping with only a concussion. She later claimed that St. Catherine had admonished her repeatedly not to jump.

Further evidence that this isn't a straightforward case of an extraordinary person inventing supernatural personages to justify her plans is Joan's own testimony that she was frightened at first by the voices and then taken aback when urged to go to war. "I answered the voice that I was a poor girl who knew nothing of riding and warfare" (a statement that is difficult to reconcile with the observations of her soldier colleagues that she seemed a born military strategist with particular skill in deploying artillery).

There were some inconsistencies: St. Michael, who had appeared to her first, failed to appear during her imprisonment, a time of great hardship for Joan. And finally her voices failed her when they assured her during her trial that she would be rescued.

This is, in broad strokes, the story of the voices. They were with Joan from beginning to end of her short but dramatic career, with few exceptions they directed her actions ("Whatever I have done that was good I have done at the bidding of my voices") and ultimately her admission that she listened to them led directly to charges of heresy, as she was claiming to be able to communicate directly with God.

There are some additional details that might be important in any retrospective attempt to understand them. The sound of church bells apparently triggered the voices. Even as a young girl Joan fell to her knees at the sound of the bells and remonstrated with the churchwarden if he failed to sound them; at her trial she testified one day that she had heard the voices three times, each corresponding with the sounding of the bells. On the other hand Joan had trouble "finding" the voices in noisy circumstances.

Perhaps the strangest of Joan's comments about her voices was her report to one of the judges at her trial that sometimes she saw multitudes of tiny angels, rather than the three saints of apparently normal dimensions. This claim, however, is from a part of the trial document called Posthumous Information and is only second-hand.

Given this information about Joan's voices and visions and the wealth of additional material about her life and exploits, can she be diagnosed (again setting aside the possibility held by some that these were truly miraculous visitations from heavenly messengers)? There has been no shortage of attempts.

As Andrew Lang pointed out in his book *The Maid of France*, published in 1909, although Joan's first experience of the visions suggested she had been entranced, there was no evidence that she was anything but fully conscious and alert when hearing her voices. Trance was not part of her experience. Lang concludes that Joan was "inspired," a vague enough term but no vaguer than suggestions by his contemporaries that Joan was somehow tapping her unconscious or "subliminal" self. Writing in the *Proceedings of the Society for Psychical Research* in 1889, Frederic Myers compared Joan's voices to the so-called "Daemon of Socrates," a voice that sometimes warned Socrates, sometimes remained silent, but apparently visited him throughout his life. Myers concluded that there was little to learn from Socrates' voice because it was a single case of a man of immeasurable genius; Joan on the other hand was not a genius, at least not in the usual sense, and so was perhaps a better example, not of madness (for neither was mad), but of "an impulse from the mind's deeper strata." (It's too bad that Sigmund Freud apparently had nothing to say about Joan: her virginity, her intolerance of coarse language and her embracing St. Catherine and St. Margaret might have made an exemplary Freudian stew.)

These notions are useful for raising the idea that the visions and voices are products of Joan's own mind, unrecognized by her as such, an idea I'll return to in a moment. They establish as well that visions per se are not an indication of any kind of pathology. That is not to say that Joan wasn't suffering from some kind of brain disorder, but she needn't have been. In Western society today visions—or hallucinations—are generally considered to be the product of a diseased mind, but there are two facts that should be weighed before applying that assumption to Joan of Arc. The first is that a surprisingly large percentage of otherwise completely "normal" people have seen visions or heard voices; the number varies according to the study but is something like 15 or 20 per cent. These can be as trivial as hearing a word or two spoken in the voice of a family member, but they can be much more elaborate than that. Hallucinations are not nearly as rare as we think. Second, cultures vary in their acceptance of visions, and Joan's world was one where visions, especially of saints, were much more common than today, and were viewed accordingly. It wasn't that insanity was unknown or unrecognized—it was—but experiencing visions of saints didn't necessarily qualify as insanity. After all, Joan's claim to visions didn't stand in the way of experienced and powerful men lining up behind her. George Bernard Shaw put it perfectly in the preface to his play *Saint Joan*:

> If Joan was mad, all Christendom was mad too; for people who believe devoutly in the existence of celestial personages are every whit as mad in that sense as the people who think they see them. Luther, when he threw his inkhorn at the devil, was no more mad than any other Augustinian monk; he had a more vivid imagination, and had perhaps eaten and slept less: that was all.

Shaw went on to remark that had Isaac Newton's imagination been of the same kind he might have hallucinated the ghost of Pythagoras walking through the orchard and explaining the fall of the apple. Joan's visions stand out as being the only peculiar aspect of her life; the rest was remarkable and heroic, but not peculiar.

Having said that, Joan was clearly not like those people who hear an incidental voice or experience even an intense one-time vision. For years her extraordinary life was guided by her voices, and several experts have been tempted, despite the opinions that her visions were an unusual manifestation of an unusual mind and nothing more, to try to identify a medical condition that might have created them.

I should begin with a medical expert's opinion of what Joan was not: psychotic. In 1984, Fred Henker, a doctor in Little Rock, Arkansas, published a report in which he applied the psychiatrist's manual of psychiatric disorders to Joan's case. According to Henker the psychoses that might apply to her would be schizophrenia (a serious mental illness involving many symptoms, the most prominent of which are disordered thoughts and auditory hallucinations), bipolar disorder (formerly called manic-depression) or a long list of personality disorders. This latter category includes a large number of conditions: some of the features fit Joan; most don't. A diagnosis of any one of them is not justified. Might she have been manic, given that she took on the leadership of an army with the most grandiose military goals? The problem here is that she showed none of the typical symptoms of talkativeness, hyperactivity, sleeplessness or (especially) distractedness.

That leaves schizophrenia, the auditory hallucinations of which suggest it might be the best fit. Indeed, there are schizophrenics who hear the voices of God or angels, although for the most part the voices are cruel or irritating, seldom providing the comfort that Joan found in hers. But more relevant are the disordered thoughts and disorganized behaviour apparent in schizophrenia. Joan was, if anything, super-organized in both thought and action.

Her behaviour at her trial provides the best examples. Faced with a panel of scholars and churchmen this unlettered teenage girl brought them to a halt again and again. When asked if she believed herself to be in a state of grace (a trick question, given that the answer "yes" would reveal that she presumed to know the mind of God, and "no" would be an admission one wouldn't want to make when being tried for heresy), Joan replied, "If I am not, God put me there, and if I am, may God keep me there!" A notary at the trial recorded that the judges were stunned. When asked if she had been present where English soldiers had been killed, she said, "In God's name yes. How gently you talk. Why don't they leave France and go back to their own country?" An English lord at the trial apparently commented, "Really, this is a fine woman. If only she was English." When asked if the angels she claimed to see were painted, Joan replied, "Yes, as they are painted in churches." She constantly amazed her interrogators with her wit and her ability to remember exactly what she had already told them. She was not schizophrenic.

With major psychoses off the list, the last best possibility for a medical explanation for Joan's visions is a discrete site of brain damage. The current favourite is one suggested in 1958 by John and Isobel Butterfield, who argued that Joan was affected by a form of tuberculosis. They began by pointing out that there is a place in the brain where sensory nerves are channelled together; any sort of localized injury there can disrupt sight or hearing. Occasionally, however, damage or disease can excite or overstimulate these nerve pathways, giving rise to what the great nineteenth-century neurologist Hughlings Jackson called "over-consciousness." He also cited several cases in which a brain tumour caused the phenomena of hearing voices and seeing faces. The Butterfields went on to point out that unusual visual events like flickering lights can trigger abnormal electrical discharges in the brain, significant, they argue, because Joan had said that she heard her voices most easily in the woods with sunlight filtering through the leaves.

They then state categorically, "All the evidence points to there having been some organic abnormality in the region of the left temporo-sphenoidal lobe of Joan's brain . . ." But while they argue that a brain tumour in that location is a good bet, partly because in young people it could grow slowly and thus allow Joan to live several years after the first symptomatic visions, they admit that the lack of any other signs of ill health is troubling. However, if instead of a growing tumour, the "organic abnormality" were a tuberculoma, the picture would make more sense. This tuberculoma would have been a hard (they use the delightful phrase "cheese-like") abscess caused by bovine tuberculosis.

Infection by these bacteria was common in the Middle Ages and Joan undoubtedly lived with tubercular cattle, so a tuberculoma is not by any means an impossibility. The Butterfields go on to try to link some of the features of tuberculosis to Joan. She apparently never menstruated, an observation that has received plenty of attention from diagnosticians. Amenorrhea can be a complication of tuberculosis, although as others have pointed out, it could have simply resulted from a combination of inadequate nutrition and late puberty. The most intriguing link, however, was that Joan's executioner reported that her heart and parts of her intestines had not burned (and were thrown into the Seine); the Butterfields contended that wouldn't be surprising if there were calcified lymph glands in her abdomen, a typical result of tuberculosis (although they didn't specifically mention the heart, calcification of that organ was not uncommon either). But even after suggesting this diagnosis, they had to admit that Joan was not a product of a disease, but a creature of courage and intelligence.

Here the scientific study of Joan takes a strange turn. The Butterfields were clearly the first to suggest the tuberculoma-in-the-brain theory for Joan's visions, and yet in 1986 the idea was put forward again—as if for the first time—by Dr. R.H. Ratnasuriya in the *Journal of the Royal Society of Medicine.* That in itself isn't so amazing—ideas can be arrived at independently—but surely if you are suggesting a medical cause for Joan of Arc's visions you would check out the possibility that someone else thought of it first. However, Ratnasuriya neither acknowledged the Butterfields nor appeared to know that the idea had been suggested nearly thirty years earlier. What is even worse is that researchers following up the idea in medical journals have quoted only Ratnasuriya—it is as if the Butterfields never existed.

However, Ratnasuriya was unable to add much of significance to the Butterfields' account other than to point out that tuberculomas in the brain were at one time very common: in the United States prior to 1900, a third of all tumours in the brain were tuberculomas, and that number was higher for young people.

One curious sidebar was that, in response to Ratnasuriya's article, a reader pointed out that calcification of the heart apparently isn't necessary for it to survive fire: poet Percy Shelley's body was burned on the beach near Viareggio and observers were startled to see the heart, as well as a few fragments of bone, did not turn into ashes.

In refining and adding to the idea of a tubercular lesion as the trigger for Joan's visions and voices, researchers have built a case for its probable location in the brain. The Butterfields argued for its being in the temporal lobe, and one of the most recent contributions to the story, a paper published in 1991 by Elizabeth Foote-Smith and Lydia Bayne of the University of California San Francisco, bolsters that argument.

The temporal lobes lie on each side of the brain—stretching from the temples to behind the ears—and damage to them can cause the condition known as temporal lobe epilepsy, or partial complex seizures. Rather than triggering violent convulsions, temporal lobe epilepsy usually causes a few brief moments of loss of contact with the world. Such seizures can be dangerous (if for in-stance the patient is driving a car) but they need not be debilitating. In the time between seizures the patient is perfectly lucid, but what is interesting with respect to Joan is that the condition often impresses an unusual set of behaviours on the person, some of which Foote-Smith and Bayne tried to link to Joan.

Intense emotion, elation and euphoria are all commonly experi-enced by people with temporal lobe seizures. Joan certainly felt joy and comfort when she heard her voices and the researchers argue that she was almost continuously exhilarated, rallying her troops in battle, constantly pushing herself and them towards her goals. Many patients come to believe divine forces are influencing them and become very religious—obviously a perfect fit with Joan. Some are extremely moral, rigidly so, and here again there is some corroborative evidence: Joan objected to any of her soldiers swearing and banned the prostitutes who habitually accompanied the troops.

There are other personality characteristics: anger, hostility, lack of humour, even altered sexuality. Joan's habit of wearing men's clothes (the subject of much questioning at her trial), her virginity and her apparent complete lack of interest in sex have all attracted the interest of analysts and biographers over the years. All of these—even transvestitism—can be associated with tem-poral lobe epilepsy. It is an impressive list, but I think you have to be very cautious about matching personality characteristics to features of an illness. The description of Joan you choose seems to depend on what you want to say about her. In contrast to the researchers in California, Bernard Shaw chose a very different list: "good-humoured . . . very pious, very temperate, very kindly . . ."

There is a precedent for attributing the acts of a person not unlike Joan to temporal lobe epilepsy. In the early 1980s an MD named Delbert Hodder, a member of the Seventh Day Adventist church, went public with his claim that Ellen White, one of the founders of that church, had suffered from temporal lobe epilepsy. Hodder pointed out that White fit all the criteria: she was intensely religious, so much so that she wrote twenty-five million words on her religious philosophy, much of which came to her in the form of visions. She was highly moral, humourless and had a strong sense of personal destiny.

But there are significant differences between Ellen White and Joan of Arc. For one thing, when White experienced a vision, she entered a trance-like state, absolutely typical, Hodder pointed out, of temporal lobe epilepsy, but unlike anything reported for Joan. White seemed to swoon for a few moments, recovering to move her arms and legs in graceful, dream-like movements; she would call out, "Glory, Glory, Glory . . . " and her eyes seemed to be focused on some distant point. Joan exhibited none of these visible signs of seizures. In addition, Ellen White had been hit in the head by a rock when she was nine years old and had nearly died. The seizures, when they began several years later, were likely centred on the injured part of her brain. Joan could have suffered from the tuberculosis necessary to create a tuberculoma in her brain, but there is no evidence she actually did.

But the real difference is in the content of the visions. White was clearly unconscious during hers, and it was Seventh Day Adventist scholars who pointed out that she seemed to have copied much of the content she claimed to have come from her visions from other nineteenth-century sources. Her visions, then, were atmospheric, not informational. Joan's were completely different: her visions might have been vague and based on standard religious iconography, but the words she reported hearing were clear, firm and to the point. They provided encouragement, advice, specific plans of action and even predictions that apparently came true. They don't sound like the visions that might accompany temporal lobe seizures, even if some aspects of her personality fit that picture.

This explanation, like so many others, considers the existence of visions, but not their content. Even if they were triggered by some brain abnormality, Joan's visions expressed lucid thoughts. Where did those thoughts come from—from Joan herself? Recent studies of schizophrenics have shown that their hallucinations likely result from their own thoughts being mistaken for the thoughts or voices of others. It might be that the little voice we hear as we're thinking (psychologists call it "inner speech") is heard by a schizophrenic as someone else's voice; it might be that the act of monitoring our thoughts as we turn them into words goes awry in schizophrenics so that they are unable to tell what belongs to them and what doesn't. Brain-imaging studies seem to be confirming that the brain "hears" a hallucination in much the same way as it hears real sounds, and that the brains of hallucinators operate differently from those who do not hallucinate. Even though Joan was not a schizophrenic, these notions about what makes a hallucination might apply to her: maybe the voices of St. Catherine, St. Michael and St. Margaret were Joan's own, unrecognized, voice (although it's curious that sometimes the voices came unbidden, while sometimes she called for them).

But even if that were true, none of the mystery and wonder would be removed, because the voices said such remarkable things. They led Joan through a planned, extended military-political campaign; they were sensible and logical. They were utterances that you would never expect from a girl like Joan, and indeed sometimes they were utterances even *she* didn't expect.

I don't think we have explained Joan of Arc yet; I doubt that we ever will. It would have been nice to transport twentieth-century technology to her and image her brain while she was hearing her voices; but even that would likely only show that her brain was perceiving voices that seemed real to her. The most we might ever be able to say is that she was, truly, extraordinary. In the words of Edward Lucie-Smith:

> Another might have heard Voices . . . but no other could have displayed her dauntless courage and gift of encouragement; her sweetness of soul; and her marvellous and victorious tenacity of will.

The Effect of Witchcraft
on the Brain

———————

———————

Every decade brings new technologies and new scientific knowledge. With them comes the temptation to try to solve—once and for all—some of the unsolved mysteries from the past. This is a different kind of science from that practised in the labs—mostly because there isn't much that can be done in the lab. It's not hypotheses backed up by experiments, at best it is informed speculation, and, as such, it changes as the fashions in science change. In the 1980s it was suggested that the Plague of Athens in 430 BC was actually toxic shock syndrome; then in the 1990s that idea was superseded by a claim that this great plague, which marked the end of the age of Pericles in Greece, was actually an early appearance of the dreaded Ebola virus. In the 1960s the monument at Stonehenge changed from a place of worship to a giant celestial computer. If you are the sort of person who wants clean, well-defined answers, this sort of science is not for you. It's never definitive, never clear-cut and certainly never complete. But it sure is entertaining. A perfect example, one of my favourites, is the attempt to understand the Salem witches.

Reading accounts of the Salem witch trials is a frightening exercise. A group of teenage girls succeeded in having twenty-two people put to death simply by accusing them of being witches. The entire Salem affair lasted only about a year but it still intrigues —and baffles—historians and psychologists. As in all cases of this kind, every age puts its own stamp on the story. In the early twentieth century, the girls' behaviour was attributed to mass hysteria. In the late forties they were put down as mischievous "bobby-soxers." Then, with perfect timing in the mid-seventies, a psychologist claimed that the girls had inadvertently consumed an LSD-like chemical, which caused them to see witch-like behaviour in others. This is the version of the Salem witches with the most curious twist, and it's worth a close look, even if there's a good chance it's not true.

Salem was a small, unremarkable Massachusetts village which was encountering difficulties during the winter of 1691-92; taxes were high, the killer smallpox was moving through the area and there were tensions between the most prominent Salem landowners. Even so, no one expected the appearance of witches. After all, the excitement over witches was dying down, and those few New Englanders who had previously been accused of witchcraft were usually released. However the villagers of Salem were Puritans, and apparently still had the capacity to believe in—and execute—witches.

The whole affair began innocently enough in December 1691, with some young Salem women apparently dabbling in the occult: trying to divine the future with a home-made crystal ball (the white of an egg floating in a glass of water) and listening to exotic tales told by Tituba, the West Indian slave of the town minister, Reverend Samuel Parris. The emotional high of these get-togethers proved to be too much for some of the girls who began to exhibit strange behaviours. Parris's daughter Elizabeth, age nine, flung a Bible across the room, an act of almost incomprehensible delinquency in the home of a Puritan minister. Her cousin Abigail Williams, who lived in the Parris home, stood up in church and loudly challenged a minister who was delivering a guest sermon. Later Abigail was seen running around in her house, hauling burning logs out of the fireplace and flinging them around the room. But she and Elizabeth were not the only ones affected. Within a few weeks, several girls were exhibiting "odd postures and antic gestures," and "uttering foolish ridiculous speeches . . ." These included twelve-year-old Ann Putnam; Susan Sheldon and Elizabeth Booth, both eighteen; Elizabeth Hubbard and Mary Walcott; and two servant girls, Mercy Lewis and Mary Warren, the eldest at twenty. They seem very young today but were referred to by the court as "grown persons."

The townspeople were at a loss to explain what had triggered this epidemic of odd and defiant behaviour, but even so it was two months before anyone publicly suggested witchcraft. Once that genie was out of the bottle, however, reluctance to admit the possibility of witchery changed abruptly and dramatically as the result of a single act. One of the girl's aunts persuaded Tituba to make a "witch's cake," a concoction of barley flour and urine from one of the afflicted girls. The idea was to feed this to a dog: in cases where bewitchment was suspected, the bewitched person would be cured if the dog shook after eating the cake. The irony here is that no one seems to know what actually happened to the dog, but the mere act of baking a witch's cake put the idea of witches in people's heads.

Town authorities, who had preferred to take a wait-and-see attitude, were under increasing pressure to do something. With the idea of witchery gaining popularity, the girls were encouraged to identify those who had bewitched them. Finally on February 29, 1692, warrants were sent out for the arrest of three women whom the teenage girls had accused of tormenting them. The three women were the slave Tituba, Sarah Osborne, an older woman who was often ill, and Sarah Good, a much-disliked town beggar.

Sarah Good's trial interrogation by the two presiding judges was indicative of what was to come:

Q: Sarah Good, what evil spirit have you familiarity with?
A: None.
Q: Have you made no contract with the devil?
A: No.
Q: Why do you hurt these children?
A: I do not hurt them. I scorn it!
Q: Whom do you employ then to do it?
A: I employ nobody.
Q: What creature do you employ then?
A: No creature—but I am falsely accused.

With this line of questioning getting nowhere, Judge Hathorne asked the girls in the courtroom to look at Sarah Good and identify her as one the tormenters. Not only did the girls assert that this was so, they underlined their claim by crying out in pain, saying they were being pinched and bitten. The court record implies that the girls' actions were not completely spontaneous, as it ends with: "Presently they were *all* tormented." Sarah Good, Sarah Osborne and Tituba were all convicted of being witches, but in the end only Sarah Good was hanged; Sarah Osborne died in jail and Tituba was eventually released.

This first trial set the tone for others which followed rapidly in the next few months. As more accused were brought to trial, the girls expanded their repertoire: soon instead of simply reacting to the presence of an accused, they described in detail how they had been tortured by the accused. Sometimes they acted in concert, a convincing demonstration of the power of the spells. In the trial of William Hobbs, Abigail Williams cried out, "He is coming to Mary Walcott!" and Mary immediately fell into a fit, thus proving that the accused was practising witchcraft right there in the court.

One of the most dramatic demonstrations was when the accused would touch the girls while they were throwing fits. If the touch calmed them (as it always did) that was a powerful indicator that the accused was a witch. This touch evidence was often coupled with so-called "spectral" evidence—seeing the ghost or spectre of the accused attacking the girls. Anne Putnam claimed in May 1692 that she had seen an apparition of two women: "They were Mr. Burroughs' first two wives and he had murdered them. And one . . . pulled aside the winding sheet and showed me the place [of the stab wound]." The unfortunate George Burroughs, a former minister in the town, was immediately the target of charges of wizardry.

The testimony against him was typically sensational. (One passerby was astonished by what he called the "hideous screech and noise" coming from inside the courtroom.) Six witnesses claimed to have seen Burroughs lift a gun at arm's length by sticking his finger in the barrel and . . . just lifting. Just before his death by hanging, Burroughs took one of the first important steps towards discrediting the case against the so-called witches. Most in attendance at the hangings believed witches were incapable of reciting even a simple prayer accurately (as if it would be unusual in that setting for anyone to stumble over memorized words). But George Burroughs went through the Lord's Prayer perfectly, apparently making the crowd restless and uneasy. But the renowned Boston minister Cotton Mather was on hand, and his tough pro-execution speech in response carried the day.

The tension wasn't limited to the court: as one of the wheels of the cart carrying a group of eight condemned witches to the gallows hill became stuck in a rut, the girls wailed that the devil was trying to rescue his servants. When one member of that ill-fated group, Samuel Wardwell, launched his appeal for mercy, he choked on smoke from the hangman's pipe. The women called out that it was the devil preventing him from speaking.

In all, thirty-one people—twenty-five women and six men—were convicted. Nineteen were hanged, two died in jail (including one of the first accused, Sarah Osborne); one, Giles Cory, was "pressed" to death (weights were piled on him in an effort to make him confess—obviously one weight too many); and other three avoided death by escaping, pleading pregnancy (the court being unwilling to take an innocent life), or, curiously, confessing. In the Salem witch affair confessing was a good thing: everyone who confessed was freed. Everyone who denied their guilt to the end was convicted.

Eventually—and gradually—reason seemed to prevail: the girl's accusations were met with increasing doubt, influential clergy raised their concerns about the possibility of convicting innocent people, and by the end of 1692 the affair had pretty well petered out. No more "witches" were hanged.

Historians of many stripes have had a field day trying to make sense of the Salem witch trial and what might have prompted a small group of young women to act as if possessed. They could have been faking it to protect themselves—as a childish prank got out of hand—but their symptoms were dramatic enough that many experts doubted that they could have been acting.

They could have been the victims of some sort of mass hysteria, although that seems not to have happened in other Puritan communities. Some analysts have argued that the affair had its roots in interfamily tensions within Salem—it is true that almost all the accusers came from one side of the town and the victims from the other. I suppose you could even argue they actually were under the spell of witches. But in 1976, psychologist Linnda Caporael took Salem witch theories into another realm entirely. She suggested the girls had been hallucinating under the influence of LSD.

But they hadn't been dropping acid in 1692. Instead, Caporael suggested in an article in *Science* called "Ergotism: The Satan Loosed in Salem?" that the young accusers had consumed rye bread contaminated with the spores of a rye-infecting fungus called ergot, the more colourful Latin name of which, *Claviceps purpurea*, translates roughly as "purple heads on a stalk." In damp weather the fungus infects stalks of rye, replacing anything from one to several grains on each of them with their own dark "sclerotium," an object that itself resembles an enormously overgrown inky purple grain of rye. The sclerotium eventually produces spores which spread the fungus.

But the chemical nature of the sclerotium is more interesting. It contains a variety of potent drugs called the ergot alkaloids, including lysergic acid amide, which has about 10 per cent of the psychoactive power of LSD. These chemicals, when consumed, can produce two dangerous forms of poisoning. One is gangrenous ergotism, which literally causes hands, feet and limbs to become gangrenous, blacken and fall off. In medieval times it was called St. Anthony's fire; fire because of the burning pain, St. Anthony because people with gangrenous ergotism sought refuge in hospitals run by the order of St. Anthony.

The second form of ergotism is convulsive, the chronic version of the acute gangrenous form. The symptoms of chronic ergotism are many, but a 1995 review of the disease in a medical journal included the following: giddiness, unusual spasms or bending of the arms and hands, maniacal excitement, hallucinations, delusional insanity and psychosis.

Ergot flourishes on rye plants when there is warm wet weather in the spring. The weather in 1691 in Salem was warm and wet in the spring, hot and stormy in the summer. There was ergot already in New England before the *Mayflower*, infesting the native wild rye. Cultivated rye was a popular crop among the settlers, with the crop being planted in the spring, harvested in August, then stored for threshing later in the fall. There is no doubt that, given the right conditions, a rye field could be infected with ergot.

Ergot infection is more likely if the rye is grown on wet lowland fields. Thomas Putnam owned one of the largest parcels of land in Salem, a swampy meadow on the west side of the village. Linnda Caporael, in her *Science* article, wondered if the Putnam farm might have been the source of the contaminated rye. She paints a picture of Putnam's ergotized rye being consumed first in the Putnam household, where three of the accusers lived, then being passed on to Samuel Parris, the minister, whose salary was partly paid in provisions (and who housed two "bewitched" girls), then on to others.

Caporael then assessed the symptoms displayed by the accusers in court in light of the possibility that they might be suffering from convulsive ergotism. Reports like Ann Putnam's "seeing" the two previous wives of George Burroughs could have been the hallucinations of ergot; feelings of being choked, pricked and pinched are similar to the reports of victims of convulsive ergotism that their skin "crawls." The fits that afflicted the girls, especially in the presence of accused witches, could have been the convulsions of ergot poisoning.

One case in particular seems a perfect fit to Caporael's argument. A man named Joseph Bayley lived outside Salem, but on his way to Boston likely spent the night at the Putnams.' As he and his wife left Salem, he reported feeling two hard blows to his chest (when there was no one else around) hallucinated the figure of a woman and felt as if he were being pinched and nipped. Of course there is no way of knowing whether he actually stayed at the

Putnams,' ate contaminated grain or actually had ergot poisoning, and it may only be one case, but it is a good one.

Caporael closes her case by pointing out that the Salem witch affair ended as abruptly as it started, the reasons for which, she argues, have never been clear. But, she points out, the next summer was hot and dry, a bad summer for ergot.

To believe Caporael's theory was to see the story of the Salem witches in a whole new light. It created an image of young women —out of their minds—experiencing horrific visions and sensations for which there was no explanation that would make any sense to anyone in Salem except that they were under the spell of witches. The fact that witchhunts were on the way out wouldn't have mattered to a village (let alone to the women themselves) frantic for an explanation. Of course, if contaminated grain was being consumed in Salem, there should have been many more than a handful of young women affected. In response, Caporael pointed out young people are affected most by ergot poisoning. Also, there were plenty of adults at the trials who corroborated the young women's testimony; perhaps these adults had been poisoned too, but not as severely.

Linnda Caporael's article appeared in *Science* in April 1976; in December that same year a lengthy reply appeared in the same journal, authored by the late Nick Spanos and Jack Gottlieb of Carleton University. They trashed Caporael's article from just about every conceivable direction. There isn't enough space to list all their criticisms, but here is a selection.

Caporael had attributed the awkward fact that most of the people living in the same houses as the accusers (and presumably eating the same rye bread) did not develop symptoms to individual differences in susceptibility to the poison. Spanos and Gottlieb quote sources who point out that it was common for all family members to be affected in epidemics of convulsive ergotism.

They also argued that many of the symptoms that should appear if this were an epidemic of convulsive ergotism, symptoms like diarrhea, vomiting and general gastrointestinal distress, simply weren't reported. The hallucinated images of people aren't even like LSD-induced hallucinations, they say, which are much more often after-images, halos, rippling surfaces—the stuff of black-light posters.

But to my mind, Spanos and Gottlieb did the most damage when they compared the symptoms exhibited by the Salem teenagers in court to symptoms typical of ergotism. The girls seemed not to be convulsing and twitching out of control as they would if diseased. Instead, they showed all the signs of reacting to cues. If an accused witch looked at them, they fell in a heap. If an accused witch bit her lip, the girls cried out that they were being bitten on their arms and hands. When Abigail Williams cried out, "He is coming to Mary Walcott!" Mary reacted immediately. If the accused touched them, their fits ceased. All of these were considered standard manifestations of witchcraft in 1692, and, in Spanos and Gottlieb's view, provide evidence that the girls were doing just what they imagined bewitched people should do.

It's interesting to note that their behaviour became better orchestrated—and more in line with what would be expected from girls tormented by witches—as time went on. First they made odd gestures and spoke out of turn. Then when witchcraft had been suggested as the cause, they started to see the spectral images of the accused; they convulsed; later again they reported being pinched and bitten. The deeper into witchcraft the community sank, the more pronounced the girls' symptoms became.

Caporael had argued that their symptoms were just too genuine, too dramatic, too *pathological* to be faked, but Spanos and Gottlieb cite a case from sixteenth-century England where people who had displayed many of the same symptoms later confessed they were faking all along.

Theirs was a hard-hitting critique, but it wasn't the last word. A professor of history at the University of Maryland, Mary Matossian, looked at both arguments and fired off some counterarguments to Spanos and Gottlieb. She pointed out that several animals had become sick, behaved strangely and died during the witch trials, and it could hardly be argued that they were acting in response to social cues. In addition, Matossian claimed that people on LSD trips are highly suggestible, countering to some degree the argument that the hallucinations reported weren't the right kind for ergot poisoning.

And that, with minor sub-arguments one way or the other, is where the story stands today. So what really happened in Salem? I lean towards Spanos and Gottlieb's argument that ergot poisoning wasn't the cause of the girls' actions. Their behaviour seems to fit too well with their claims of being tormented. There were even whispers at the time that the girls were playing games, games that likely got out of their control. On the other hand, there are unexplainable cases, like the man thumped in the chest, Joseph Bayley, and the *mother* of Ann Putnam, who came down with the same kinds of symptoms as her daughter early on in the affair, only to recover after having appropriate Biblical passages read to her.

Perhaps some sort of combination is the real answer: ergot poisoning that triggered some of the early strange behaviour, the antic movements and odd outbursts, followed by the human need for explanation and a ready one supplied by whispers of witchcraft. Maybe unexplained events—a mother showing the same symptoms as her daughter, a man feeling a thump on the chest—happen all the time but are ignored. Against a background of suspected witchcraft, however, heightened awareness seizes on such events and attributes meaning to them. Whether it was poisoning, politics or pretense, the most chilling aspect of the Salem witch affair is the eloquence of innocent people on the gallows hill, just before their death.

The Effect of Witchcraft on the Brain

Sarah Good, one of the first accused, a beggar and extremely unpopular in Salem because of her unpleasant manner, went to the gallows on July 19th. When she was asked to confess by the assistant minister of the Salem church, she said, "I am no more a witch than you are a wizard, and if you take away my life, God will give you blood to drink."

The Vinland Map

On October 11, 1965, Yale University Press announced the publication of a scholarly book with the peculiar title, *The Vinland Map and the Tartar Relation*. (The odd title of the book referred to two medieval documents—bound within one cover—that Yale University had purchased; the book itself was a technical analysis of the two.) The publisher had originally planned to release the book on Saturday, October 9th, Leif Erikson Day, but had to change the publication date to the following Monday, the 11th. Little did they know the furore that would greet that innocent decision. What was intended to be a modest academic celebration turned into chaos, as the university and the authors of the book were vilified, accused of a direct attack on that cornerstone of American history, Christopher Columbus's discovery of the New World on October 12, 1492. The *Chicago Tribune* called it "The Map That Spoiled Columbus Day."

Why the indignation? One of the two medieval documents, the

The Vinland Map

Vinland Map, portrayed the world including Iceland, Greenland and "Vinland," a large island that scholars were sure represented North America. The island was divided into three by two deep inlets (possibly Hudson Strait and the St. Lawrence River); the three parts might well be the Helluland, Markland and Vinland described by the Norse sagas dating back to AD 1000. Not only that: the experts agreed that the map had been drawn around the year 1440. This map was the first ever found to provide solid evidence that Norse explorers like Leif Erikson were actually the first people to "discover" America, and they did it nearly five centuries before Columbus. No wonder there were outraged headlines.

Of course that was 1965. There wouldn't be nearly the same fuss now—the Norse settlement discovered since at L'Anse-aux-Meadows in Newfoundland proves that the Norse were here, map or no map, and even if that weren't the case, the 1992 celebration—and protest—of Christopher Columbus proved that there is a new awareness of the other side of his great discovery, the importation of disease, violence and intolerance towards Native peoples in the Americas.

But while Americans and Canadians have largely forgotten the Vinland Map brouhaha, the experts haven't, for they continue to fan the flames of a controversy of their own: is the map a fake? The arguments on both sides offer a glimpse into two fascinating worlds: that of medieval historians trying to work from circumstantial evidence and scientists trying to apply the highest technologies available, both groups attempting to authenticate the map—or not. It might seem at first glance that the scientific approach would be more objective, but, as the evidence is rolled out, the two have more in common than you would think. When it comes to science, it's not so much the technology that counts—it's who is using it. And while a majority seems to have decided that the map is the real thing, there are still nagging doubts.

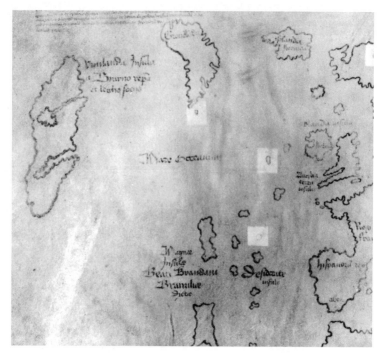

The island on the left of this map is marked Vinland. Is this the earliest image of North America, drawn from Viking legends? Or is it instead one of the cleverest fakes ever?

Doubts that, maybe surprisingly, haven't been resolved by the hard-nosed technological approach.

Obviously the discovery of the first map ever to depict North America would have been controversial no matter what, but the way the Vinland Map came to light gave it an extra sheen of suspicion. There are many twists and turns, shadowy book collectors, dealers and unidentified characters involved, but the essentials are this: in 1957 a New Haven, Connecticut, bookseller named Laurence Witten II, while in Europe buying manuscripts, purchased an unusual two-part document. The first two pages (actually a single piece of vellum folded down the middle) showed a map of the world, the Vinland Map; it was bound to a sixteen-page document in Latin describing a trip to Asia in 1245-47 by a Roman Catholic friar named John de Plano Carpini. This trip was an attempt to make contact with the empire of the Khans, specifically to meet (and check out) the Great Khan, Kuyuk, grandson of Ghengis Khan. It was a harrowing journey of nearly 13,000 kilometres, mostly on horseback, under constant threat of starvation or death at the hands of the Tartars. This manuscript, essentially an intelligence report on the Tartars, is called the *Tartar Relation.*

Witten bought the pair for US$3500, and while it was later called "the greatest bargain in all the world" he was in fact putting a lot on the line. At first glance the *Tartar Relation* looked like the real thing, but the map was a different story. Even though it was bound to the *Tartar Relation*, it was clear that hadn't always been the case. The binding looked to be twentieth century, and even the wormholes in the two documents didn't match. Also there was no record of its origin. These inconsistencies opened up the real possibility that the map was a fake, a recent forgery. Potential buyers before Witten had sought opinion from experts at the British Museum who said that although they believed it to be authentic, they wouldn't say so unequivocally—the map was too spectacular and would likely be too controversial. However Witten took the plunge anyway, mostly because of the handwriting. It seemed to be the same hand on both the map and the *Tartar Relation*, and given that there were sixty-seven inscriptions on the map—in sometimes nearly microscopic script—and 1450 lines in the Tartar Relation, Witten thought forgery, or at least convincing forgery, was out of the question.

When Witten returned home to New Haven, he immediately told Tom Marston, curator of classics at Yale, about his new purchase. However, they agreed that exciting as it was, nothing much could be done about the map until there was better evidence of its authenticity. Little did they imagine at the time where that evidence would come from. More than a year passed before the next remarkable, almost unbelievable event in the story. One winter afternoon Marston called Witten to suggest he look at some new manuscripts the curator had bought. One of them was a worn fifteenth-century sheepskin binding containing a section of Vincent of Beauvais's *Speculum Historiale,* a not uncommon but very lengthy medieval encyclopaedia and history of the world. Witten noted that the writing looked something like his Vinland Map and asked Marston if he could borrow it overnight. After dinner he began to compare the two, and immediately saw the style of writing, the place and time of origin seemed to be similar. So were the watermarks. Amazingly, the pages were exactly the same size too. Witten describes what happened next:

> My heart began to pound but there were more surprises to come. After some additional scrutiny my adrenaline began to flow as it dawned on me that the wormholes through the Vinland map exactly matched those at the front of Vincent de Beauvais's *Speculum,* while the ones in the *Tartar Relation* just as perfectly matched those at the end of the same volume.

No wonder the wormholes in the map and the *Tartar Relation* were mismatched—this other manuscript, the *Speculum*, just purchased by Tom Marston, had originally been sandwiched in between. At some recent point they had become separated and the Vinland Map and the *Tartar Relation* had been reassembled within their twentieth-century binding. The odds against all three showing up in New Haven, Connecticut, in the late 1950s seemed absolutely incredible (and in fact led at least one skeptic to suggest that the amazing coincidence had been arranged by unscrupulous—but clever—dealers in Europe who knew the map to be a forgery).

But skepticism was rare at this early point, for several reasons. Smudged traces of writing left on the covers of the *Speculum* binding suggested it was prepared during a great church gathering called the Council of Basel that ran from 1431 to 1449, while independent investigation of the paper showed it had been made at a mill in Basel in 1440 or 1441. There was even a one-line inscription on an otherwise blank page that, although mysterious before, could now be understood to be referring to the first, second and third parts of the speculum. What better proof could there be that the three documents had once been bound together as one? And because there was no doubting the authenticity of either the *Speculum* or the *Tartar Relation*, things looked pretty good for the Vinland Map.

Pretty good but not perfect. One stumbling block was the uncertain history of the documents. Laurence Witten had given his word to the previous owner not to reveal his name (for tax reasons), a promise he kept even when the controversy was at its height. All Witten could say was that the previous owner was sure that the documents had been in his family library for at least two generations. But these were vague assurances from an unknown person—certainly no guarantee that fraud of some kind was not involved. A one-of-a-kind map with an uncertain past: the perfect subject for experts to strut their stuff, and in the ten years following the 1965 revelation of the map, that's exactly what they did.

The first wave of experts included historians, cartographers, geographers and linguists, all trying either to establish the reasonableness of the Vinland Map or to hint at the possibility of fraud. Did it make sense that some unknown individual attending the Council of Basel would have had access to information about the existence of North America; could a map produced by his hand have made it to New Haven, Connecticut, 517 years later?

There wasn't much doubt that the Norse had visited North

America a thousand years ago. Even though they lacked the navigational instruments used at the time by other Europeans, they were obviously accomplished sailors. According to the sagas and a variety of written histories, Erik the Red led the settlement of Greenland in the 980s, apparently at a time when temperatures in the North Atlantic were much warmer than today and sea ice much less of a threat. Those Greenland colonies flourished for centuries until they finally declined and failed because of worsening climate. As far as North American contact is concerned, the best guess is that about the year 986 an Icelander named Bjarni Herjolfsson was the first actually to see the coast of North America after being blown off course on his way from Iceland to Greenland (the fact that he didn't bother going ashore was later judged by his peers to have been pretty stupid), while Leif Erikson is likely the first to have landed, around the year AD 1000. Erikson's "Helluland" is assumed by most to be the southern part of Baffin Island, "Markland" Labrador, and Vinland the northern peninsula of Newfoundland. However, even though Leif spent a winter in Vinland, it seems as if the Norse failed to establish much of a permanent presence in North America after that, even though there were apparently occasional visits for another four hundred years. However this handful of visits to North America was well enough recorded that it seems completely reasonable that Europeans in the 1400s might have been aware they had happened.

That of course doesn't prove the Vinland Map is the real thing. The experts also devoted a lot of their time and energy to the actual depiction of the North Atlantic, especially Greenland and Vinland, to see if, in their details, there might not be hints of twentieth-, not fourteenth-century knowledge.

Ironically, although the public reaction centred on Vinland, the experts were fascinated by Greenland, because it was shown as an island. World maps previous to the 1440s had all portrayed

Greenland as the tail-end of a huge arching peninsula beginning in northern Europe and extending over the North Atlantic. Did the Vinland Map version suggest that the Norse had sailed completely around Greenland, mapping it as they went and so proving it had no connection to Europe? It would have to have been by sail, because the Norse settlements didn't extend nearly far enough north on either the east or west coast to make Greenland's island nature apparent. Could they have sailed around the polar end? A good question, because circumnavigation of Greenland in modern times is extremely difficult most of the time—much of it is buried under ice. We know from the Norse accounts that sea ice was less troubling to navigation at the time they were exploring the North Atlantic than it is today, but even then conditions were beginning to degrade as the climate cooled: by Leif Erikson's time, what had been a straight sail westward from Iceland to the southern tip of Greenland had to be changed to an initial southwest leg to avoid ice, then turning northwest to hit land. If circumnavigating Greenland presents challenges to modern icebreakers, could the Norse have done it with their serviceable but relatively flimsy seagoing vessels? And if the Vikings really did make their way around the coast of Greenland, why isn't Ellesmere Island on the map? You can't miss it if you're sailing down the west coast of Greenland. Well, said some, maybe they didn't have to actually sail around it—they could have learned Greenland was an island from the Inuit. This raised all kinds of debate about whether the Norse would have even talked to the Inuit, and even if they had discussed the shape of Greenland, how easily could that verbal information have been translated into an outline on a Norse map that then could have been copied on the Vinland Map? Or was it an Inuit map that was copied?

There is even a third possibility. Cartographers of the time were touched by islomania: they often depicted little-known lands as islands as a matter of convention. The creator of the Vinland Map might have treated the largely mysterious Greenland in that way, just as he depicted Vinland as a large island divided roughly into three by two huge inlets on the east side. One of the only available ways of testing these different theories was to compare the shape of Greenland on the Vinland Map to its shape on a modern atlas—an extremely accurate rendition would support the claim that the Norse had seen the entire island.

In the 1965 edition of *The Vinland Map and the Tartar Relation*, on page 184, modern Greenland is displayed beside Vinland's Greenland, and they are a surprisingly good match. The general shape of the coasts are the same, the Vinland version exhibits some of the right zigs and zags at the appropriate places, and a dozen specific fjords and stretches of coastline can be identified in both. However, in the thirty-plus years since that edition, expert opinion has changed. In 1995 Yale University Press published an updated version of *The Vinland Map and the Tartar Relation*, and in one of the new introductory chapters, George Painter of the British Museum, a man who has believed the Vinland Map to be authentic from the beginning, nonetheless admits that the comparison of the two Greenlands in the original edition of the book is misleading. For one thing, on the Vinland Map, Greenland is only about as long as a matchstick—a little over three centimetres. That makes it much harder to buy the idea that each little fjord and inlet was put there purposefully. Painter thinks there is no true correlation between reality and the map and that the most likely explanation is that the map's creator adopted a standard medieval technique that could be called "when in doubt, make the shore crinkly." Painter doesn't think the Greenland coast is any different —or any more realistic—than any of the other coasts on the map, at least when it comes to such fine detail.

Greenland was but one example, but a perfect one, of how

speculation, even of the learned kind, falls short of answering what was the crucial question from the beginning: Had someone faked the Vinland Map? Scenarios were put forward as to how a forger could have done it: he (or she) uses a blank parchment sheet from the two authentic documents, the *Speculum Historiale* and the *Tartar Relation*, to create the map (using medieval-style ink), then while all three are still bound, creates the fake wormholes that penetrate all three. (There are dastardly types who would create "worm" holes using tiny drills the same diameter as a bookworm.) Then the forger separates the documents, binds the map and the *Tartar Relation* together and puts them on the open market, holding the *Speculum* back for a brief period, confident that when it is released it will—sooner rather than later—join the other two, in what will look like a dramatic and nearly impossible coincidence. It was possible with each of those twists and turns in the story to object, to point out how unlikely each was, to argue that fit was just too good. But the uncertainty, the *doubt*, wouldn't go away.

So in February 1972, Yale turned to science for something more definitive. Had they known how many more twists and turns lay ahead they might have had second thoughts. An independent scientist named Walter McCrone was chosen to analyze the map. McCrone had built a reputation as a microscopist supreme; his six-volume work *The Particle Atlas* is a standard reference on the identification and analysis of microscopic particulate matter of all kinds. McCrone assigned one of his colleagues to pick fifty-four extremely tiny particles from the surface of the Vinland Map (tiny enough that all fifty-four together weighed less than a millionth of a gram and if piled together would be barely visible to the naked eye) and examined them with an array of technologies: microscopes, ion microprobes, X-rays and electron diffraction, all aimed at determining exactly what substances were present on the surface of the map. His conclusion wasn't long in coming and it was a shocker: McCrone's analysis showed that the ink that had been used to draw the map and write the inscriptions contained very large amounts of titanium—in some places as much as 50 per cent.

Titanium is a metal that when combined with oxygen forms a titanium oxide called anatase, used widely today as a white pigment. It seemed extremely unlikely that a fifteenth-century ink could contain as much as 50 per cent titanium. But the clincher came when these pigment particles (so small that it would take one hundred thousand of them to span your little fingernail) were examined with the electron microscope. It revealed the crystals to be round and regular, typical of those created in industrial processes first used to manufacture titanium pigments in the 1920s. The only one who saw any humour in the results was McCrone himself, who said that the chances of a five-hundred-year-old map containing such pigment globules were about the same as Admiral Nelson's flagship at Trafalgar being a hovercraft.

The one loose end was the fact that the ink on the map was yellow, but anatase is a brilliant white. However, when anatase was first being produced eighty years ago it was contaminated with iron, giving it a yellowish hue, perfect for imitating faded medieval ink. McCrone's report was devastating. Most of the world at large and many of the experts were convinced that the suspicions had been borne out, the chemistry didn't lie and the Vinland Map was a brilliant, imaginative, scholarly and almost perfectly executed hoax.

However, a few insiders clung stubbornly to their belief that the map fit just too perfectly into the scenario that had been created for it to be dismissed by the first science that came along. And indeed there were some puzzles in McCrone's data. For one thing, there were places where there seemed to be ink but no titanium. Some of the faithful—scientists among them—were uncomfortable with the idea of extrapolating from a millionth of a gram of material to the entire map, an unavoidable consequence of picking micro-particles from the map's surface.

Perhaps inevitably, a second scientific analysis was performed on the Vinland Map in January 1985, at the University of California, Davis. There, a group led by Tom Cahill used particle-induced X-ray emission, PIXE, to catalogue the chemicals present on the Vinland Map. Cahill's team at Davis has analysed more than a thousand ancient documents, including a Gutenberg Bible and the Dead Sea Scrolls. In PIXE, a high-speed beam of protons, particles from atomic nuclei, is aimed at the target—in this case the ink line on the map. The protons dislodge electrons from the ink atoms, forcing these atoms to rearrange their remaining electrons. In doing so all chemicals emit a unique set of X-rays. So you simply aim the beam at the ink, read the X-rays coming back and you have a list of the chemicals present.

The results were stunning. PIXE detected levels of titanium tens to hundreds of thousands of times lower than those reported by McCrone. Cahill reported seeing no crystals of any kind, let alone of the twentieth-century compound anatase, and he and his colleagues went even further. They drew fake map lines using a modern titanium-based ink on a sixteenth-century piece of parchment. Then they erased those lines to the point where they were no longer visible to the naked eye. Even so they still found levels of titanium twenty thousand times higher than they had detected on the Vinland Map. The Davis group had even found more titanium in the ink of their Gutenberg Bible than they found on the Vinland Map. If PIXE were capable of finding titanium in a fake ink even when it can't be seen, yet detected none on the Vinland Map, how could McCrone have found so much?

The two results cannot be reconciled. Tom Cahill has put it well: "Let's say there's a piece of ink and he [McCrone] pulls a crystal off it, and it's fifty per cent anatase. He extrapolates to say the whole ink is fifty per cent anatase. Because we analyzed all the ink present, we have to extrapolate nothing. We found the ink was highly variable across the map: some of it has titanium, some doesn't." McCrone contends that by surveying broad areas, PIXE reduces the concentrations of titanium to insignificant levels. As to how McCrone is able even to find modern-looking crystals with so much titanium in them, Cahill raised the possibility of contamination—a white-painted room is full of such crystals, many of them airborne.

That was the situation in the mid-1980s. Since then Walter McCrone's analysis has taken some further hits. Two of the new introductory chapters in the rereleased version of *The Vinland Map and the Tartar Relation* include thinly disguised or undisguised attacks on McCrone. One is by George Painter, who attacks anyone and everyone who has questioned the map's authenticity and characterizes McCrone's theories of how a forger might have executed the map as "absurd," "pointless," "incredible" and "preposterous." Two chapters later, Tom Cahill, the scientist behind PIXE, adds these comments about the validity of the McCrone procedure: "we can find no evidence that the critical particle removal process was guided on site by manuscript experts as it was being performed," and goes on to lament the "lack of prior experience of the McCrone investigators."

While Cahill suggests that the titanium that Walter McCrone found in crystals from the Vinland Map might be a modern contaminant, a different explanation (and just as convincing) has come from Jacqueline Olin of the Smithsonian Institution. She has shown that minute titanium crystals, of approximately the type seen under McCrone's microscope, can result from a typically medieval preparation of ink, which involved roasting materials until they had been reduced to fine powders. Olin claims that some anatase-like crystals, modern in appearance, could result from this process if the furnaces used reached high enough temperatures.

For his part, McCrone, on the McCrone Research Institute website (www.mcri.org), scoffs at the idea that the necessary heat could have been generated in a furnace being used by a fifteenth-century ink-maker. He also wonders why, if his method is so flawed, he and his colleagues found two hundred times as much titanium (in modern form) on the Vinland Map as they did on either the *Tartar Relation* or the *Speculum Historiale.* If all three documents were prepared at the same time using similar or even identical inks, why is the map so much richer in titanium? And McCrone, too, can boast of support from the Smithsonian. Kenneth Towe, in that museum's department of paleobiology, agrees with McCrone that the likelihood that a fifteenth-century ink-maker could produce true anatase crystals is exceedingly small and that in fact Olin's crystals aren't like the ones on the map. He has also analyzed Cahill's data statistically and concluded that the amounts of titanium in the ink are significantly higher than those from the parchment alone. If indeed the titanium in modern crystalline form has drifted onto the map from white-painted walls, why did those particles land only on the inked portions of the map?

There is an object lesson here for anyone out there who still believes that science is an unbiased route to the truth. In this case each individual scientific approach leads to a version of the truth, but the two independent approaches are diametrically opposed. Walter McCrone has a reputation for skepticism—he performed a similar particle analysis on the Shroud of Turin and demonstrated (to the satisfaction of many) that the mysterious image on the shroud had been painted. It's not surprising he concluded the map to be a forgery. On the other hand, would Tom Cahill have gone to the trouble of retesting the Vinland Map if he didn't believe there was still a good chance it was authentic? I'm not suggesting that either scientist has been dishonest, but it's a very good bet they started with different preconceptions and after all, the answers you get are constrained by the questions you ask. At any rate, in the case of the Vinland Map, straightforward scientific approaches cannot, by themselves, answer the question everyone is asking. Even if McCrone had found no titanium, or Cahill had found plenty of it, there would still be doubts. The map could be radio-carbon dated, found to be five hundred years old and still be a forgery.

After endless arguments over the heat of fifteenth-century furnaces or the motives of a putative forger choosing a 1920s house paint to fake a fifteenth century map you can't help conclude that belief, not science, is what is most important here. And in that sense the saga of the Vinland Map is no different from any other kind of science. Until the data is overwhelmingly persuasive, belief holds sway.

As one last thought, assume the Vinland Map is real. Imagine the reaction of the man who laboured so carefully on its precise coastlines and near-microscopic inscriptions, if he were told that five hundred years in the future, scholars would be holding confer-ences, blowing up his images to thousands of times their actual size and bombarding the map with mysterious streams of particles, all in an effort to determine if he really had drawn it. I think there might be a smile on his lips.

The Burning Mirrors of Syracuse

———————

———————

The caricature of the nerdy scientist in his/her lab coat, complete with pocket protector, uttering incomprehensible jargon is bad enough. But the implied character of the person behind the wardrobe is worse: strait-jacketed by conservatism, too quick to demand hard data, hell-bent on reducing life's mysteries to uninteresting sets of numbers and graphs. Mysteries aren't *reduced* by science—they take their place in grander landscapes. They are made elegant. The truth is that scientists love a mystery as much as anyone (it's their business to chase mysteries after all) even when, as in this case, there is almost no chance it will be solved. Why? Because it's intriguing, challenging and fun.

Spring, 213 BC. The hot sun sparkles off the waters of the great harbour at Syracuse, the port city on the eastern coast of the island of Sicily. Rome and Carthage are engaged in the Second Punic War: at this very moment the brilliant Carthaginian general, Hannibal, is roaming freely round Italy with his troops, the

Romans reluctant to engage him, knowing all too well his superior military skills. At the same time, Hannibal realizes that to attack Rome outright would be folly, and so he remains on the move.

One of Hannibal's main problems, and one of the reasons for his deliberation, is that he is remote from Carthage (on the north coast of Africa in what is now Tunisia) and the supplies he needs to maintain his army. That accounts for Sicily's importance: if Carthage controls it, a pipeline to Hannibal is then easily maintained from north Africa through the island to the Italian coast. If Rome supplants Carthage in Sicily, Hannibal is pretty much on his own. In 213 Carthage has more or less locked up the south half of the island, but the Romans control key ports on the north side. Syracuse is a Greek outpost whose allegiance is to the Carthaginians.

The Romans, under one of their best consuls, Claudius Marcellus, have taken those Sicilian towns they can and are now turning their attention directly to Syracuse. Marcellus, ruthless and fearless, has gathered together a fleet of sixty ships, Roman quinqueremes, each with 150 rowers on three decks, five to an oar, plus seventy-five soldiers, twenty-five sailors and officers. Eight of the ships are lashed together in pairs, each pair supporting between them an ingenious siege engine called the sambuca, a huge pivoting ladder with a soldiers' compartment at the upper end counterbalanced by two and a half tons of rocks at the lower. Because it is counterbalanced, the sambuca can reach city walls from much greater distances than ordinary siege ladders; it is also closed in, so the soldiers in the compartment aren't exposed to enemy fire as they approach.

On this sunny day the Roman fleet approaches the sheer walls of the Syracuse harbour from the south. They know this is not likely to be just another encounter. The greatest mathematician of the time, the legendary Archimedes, lives within those walls. And although mathematics is his life, and the only intellectual pursuit he values, he has made his reputation not by theorems, but by the design of war machines. As with Leonardo seventeen centuries later, weaponry pays the way and makes possible the free time to write about conic sections and hydrostatics.

As the rowers manoeuvre their ships close to the walls, the assault begins. Huge boulders are flung onto the ships by giant Archimedean catapults. Those ships that evade the boulders and move closer then run into a hail of smaller projectiles fired through loopholes in the city walls. Even so, some ships of the fleet edge closer and begin to deploy the sambucae. In response, long wooden arms reach over the walls and drop massive chunks of lead onto the ladders and their ships. Then the rowers are stunned by the appearance of strange never-before-seen weapons. Giant cranes swing out from behind the walls, dangling grappling hooks down onto the ships. If the hooks happen to catch the prow of the ship, the crane is suddenly levered up (Archimedes did, after all, claim of his levers, "Give me a place to stand and I can move the earth") and the ship is jerked into a vertical position, standing on its stern in the water. Then the hooks are released, sending it crashing back. Marcellus, though no humourist, was moved to remark that Archimedes "uses our ships to ladle water from the sea."

The sailors are so terrified that any slight movement on the other side of the city walls throws them into a panic, shouting, "Archimedes is going to set one of his engines against us." But what they have experienced so far is nothing to what is coming. As those ships still seaworthy circle slowly a bowshot from the wall, a crowd of Syracusans, arranged as if in a choir on a hillside above the walls, move their arms suddenly in unison and a shatteringly luminous beam of white light appears out of nowhere, focused on one of the nearest ships. Moments later the vessel bursts into flame. The beam sweeps across the water to another ship. It, too, ignites. The Roman fleet is now in complete disarray, and Marcellus calls a retreat. Archimedes has proven once again he is a genius of geniuses and has saved Syracuse, at least for the time being.

Legend tells us that is how it would have looked that day in the Mediterranean. Archimedes had trained his men to hold mirrors in front of them to reflect the sun onto Roman ships one by one. The sun's rays, so concentrated, set the wooden ships on fire almost immediately. It would be difficult not to be intrigued by this story of a solitary genius twenty-two centuries ago using science and technology well ahead of its time to thwart the military might of Rome. However there is a good chance that it's a great story but not a true story. It has both historical and technological problems.

There is no doubt that Archimedes' weapons were an important part of the defences of Syracuse against the Roman attack. From all believable accounts, there were cranes flipping quinquiremes on their ends and Romans being crushed by lead or boulders. It is also certain that the attack was repulsed. All of this we know from the best historical accounts of the battle. But none of the three main sources, Livy, Plutarch or Polybius, say anything at all about mirrors or even fire, and there doesn't seem to be any reason why they wouldn't have if it had happened. Even more damning is the fact that Plutarch was familiar with the concept of burning mirrors: he described them in some of his other writings.

So where does the story come from? The first reference to it appears in the writings of Lucian in the late second century AD, about four hundred years later. And Lucian isn't specific enough; he writes that Archimedes "burned the ships of the enemy by means of his science" which might merely be a reference to flinging fire pots —mixtures of flaming pitch, tallow and sulphur—onto the decks of ships.

The first really good reference to burning mirrors is much later than that, seven hundred years after it was supposed to have happened, by a writer named Anthemius. Even then, he wrote of the "unanimous tradition that Archimedes used burning mirrors to burn the enemy fleet a bowshot off." He made it clear that this "tradition" was maintained by several other authors, but we have no idea who they were. And if you distrust the veracity of a several-hundred-year-old tradition, then you have to wait *another* seven hundred years for accounts of the events in the harbour at Syracuse which claim to be based—at least indirectly—on actual witness testimony. But these accounts are written by two undistinguished historians (one of whom is known to have made up a similar story about burning mirrors in the harbour of Constantinople) and are based on testimony that either has been lost, or worse, never existed.

Most historians faced with that sorry line-up would have abandoned ship a long time ago. But the story of the burning mirrors is so compelling—and the reputation and demonstrated abilities of Archimedes so great—that the story lives on. In fact it has inspired many scientists, at least one of them certifiably brilliant, to investigate whether or not Archimedes could have done it.

There are actually two questions here: did Archimedes know enough about optics to design mirrors that could focus the sun's rays, and second, could such an instrument be put into actual military practice.

It appears that there wasn't a lot that Archimedes didn't know. As is often the case with great scientists (Newton and the apple), he has been reduced over the centuries to a caricature, a man running naked through the streets of Syracuse yelling "Eureka." (He was moved to do that because he had figured out from his bath how to use the displacement of water to calculate whether a gold crown belonging to his friend King Hiero was actually a fraudulent mix of gold and silver.) But Archimedes was a brilliant mathematician and engineer, and one of his specialities was optics, the properties of lenses. Designing and building burning mirrors was not a trivial problem. A single parabolic mirror, the regularly curved shape necessary to bring the sun's rays to a focus, isn't practical. The technology didn't exist to make a mirror—or any kind of reflecting surface—of the necessary size and precise shape. A rough parabola made up of many single flat mirrors would do the trick, but those individual mirrors would have to be individually adjustable.

Archimedes knew that a good approximation of a parabola could be made from flat surfaces—no problem there. And while it's not clear from his writings whether he understood how to focus the sun's rays using a parabola, there are those throughout history who have had the faith he knew enough of the related geometry to have figured it out; it's just that those particular works of his have been lost. Remember Archimedes was held in high esteem by his colleagues (and those who followed) for his mathematics, not his military engineering.

Yet there are nagging doubts about this too. You'd think a work written a couple of decades later called "On Burning Mirrors" might mention this fantastic event, if not highlight it. But this manuscript by Diocles, which only came to light in the mid 1970s, says nothing at all about Archimedes and burning mirrors, even though the author clearly knew who Archimedes was. So we can never know exactly what Archimedes knew, nor what valuable writings of his have been lost, especially when the Romans eventually took Syracuse and sacked it, killing Archimedes in the process.

Knowing the theory of burning mirrors doesn't guarantee that one which works can be built, and it is this question that a variety of scientists have tried to answer over the centuries. Could Archimedes have set Roman ships on fire?

The most impressive evidence that it might have happened was provided by a legendary eighteenth-century biologist, Georges

Louis LeClerc de Buffon. Buffon's main claim to fame was a forty-four-volume Natural History that as a young man he had set out to make his life's work. But he was much more than a studious scientist: wealthy, handsome, a great dresser, a terrific writer, a man about Paris. It's not surprising that revolutionaries broke into his tomb, scattered his bones and guillotined his son.

Voltaire called him a second Archimedes. I'm not sure of the context of that remark, but it may have had something to do with Buffon's intense interest in the idea of the burning mirrors; he actually made his own set and tested them.

There is one really difficult problem in creating a burning mirror: because the sun is a disc (appearing to be about the size of an Aspirin held at arm's length), the reflection from any point on a mirror is always a disc, and the further away the target, the broader and less intense the reflected disc is. This problem can be overcome by a concave or parabolic mirror because it focuses many thousands of different disc-like sun images at the same place and so creates a huge amount of heat. Even so, such a mirror focuses the light at only one specific distance, making it nearly useless for combat. Ships bob up and down and back and forth, and the focused beam must be maintained on the same spot on the ship's hull long enough to set it on fire.

Even in Buffon's time, a huge one-piece parabolic reflector was too difficult to build, but a myriad of flat mirrors could be fitted together as a mosaic equivalent. That is what Buffon did. From small-scale preliminary experiments he calculated that an assemblage of small mirrors ten metres across should be able to ignite wood at a distance of about eighty metres. He settled on 168 mirrors each fifteen centimetres wide and twenty centimetres high. Of course as soon as he had completed it, one cloudy day followed another, but on April 10, 1747, the sun shone and his wall of mirrors set a piece of wood fifty metres away on fire. It was enough to convince Buffon that Archimedes could have done it—after all, the Roman ships had been said to be no more than a bowshot away from the city walls, which probably amounts to about thirty metres.

However, this demonstration hasn't convinced every skeptic. Even though the wood ignited in this instance (and again in a Greek experiment in the 1970s where each of about fifty sailors focused a polished shield-sized mirror on a wooden dinghy anchored offshore), did it continue to burn? Did it ignite quickly enough to be a useful weapon? What about the obvious fact that no Roman ship would have remained still for the estimated minimum of thirty seconds needed to set the ship on fire? How much harder would it be to ignite wet wood than dry? And how would each of the Syracusans holding a mirror know exactly where to aim it so that all the individual beams came together in one place?

The noted futurist and science fiction writer Arthur C. Clarke offered one solution to this last question in his short story "A Slight Case of Sunstroke." The leaders of a soccer-mad country seek to ensure victory in a game with their most-hated international rival by handing out silvered souvenir programmes at the beginning of the game. At the appropriate moment, when the dishonest referee has just robbed the home team of a legitimate goal, 50,000 soldiers positioned on the sunny side of the stadium lift their programmes and reflect the sun at the hapless referee. He vaporizes instantly in the heat. The moral of this story is, enough soldiers aiming mirrors should ensure a sufficient number of "hits." Some engineers have picked up on this idea and suggested that Archimedes' soldiers could have had tiny holes drilled through their mirrors through which they could sight their target.

One of the most recent analyses of the Archimedean version of the problem calculated that given the height of the sun in the sky in the spring of 213 BC, the direction from which the Roman fleet was arriving, even the material out of which reflectors could have been made (silver best, copper pretty good), there would have had to be 420 men on shore each holding a piece of metal about the size of a card table. And even at that there's little chance they could have produced an explosive beam of reflected sunlight.

However, these same calculations showed that a much smaller force, say fifty men, could direct enough sunlight onto the quarterdeck of a ship fifty metres away to burn severely the steersman and officers standing there and create substantial panic. Perhaps it was fear, not fire, that forced the Roman fleet to retreat.

I think the evidence is pretty good that we will never solve this particular mystery. It's very unlikely, although not impossible, that some new source describing those events of twenty-two centuries ago will come to light. We're left with the ambiguous historical material and the not-quite-definitive experiments. But does it really matter that we will never know? The pursuit of the mystery is at least as exciting as the solution and has the additional benefit that it can go on forever.

One last thought: as this issue was being rehashed in the pages of *Applied Optics* in the 1970s, one attentive reader pointed out that vacuum metallized glass was in common use in the windows of high-rise buildings. Metals such as gold, aluminum and chromium applied in this way have fantastic reflective capacities. The reader noted that a hotel in the American south which had a curved south-facing wing could, had its windows been coated in this way, have generated a fantastic amount of energy right where the hotel swimming pool was located. As he put it, "One imagines a diver stepping out on the springboard and going up in smoke!"

The Monks Who Saw
the Moon Split Open

———————

———————

This chapter is unlike any of the other mysteries in this section in that it might actually be solved one day.

On the evening of June 18, 1178, five anonymous individuals in England were witness to a spectacular event in the night sky. While it is true that we might never know exactly what it was they saw, surprisingly, clues continue to accumulate today—more than eight hundred years later. The most spectacular possibility is that these men actually saw the creation of a crater on the moon.

There is a nicely detailed record of what they saw in the *Chronicles of Gervase*, a monk at Christ Church Cathedral, Canterbury, England. Here is the English version of his description of a very strange series of events:

This year, on the Sunday before the Birth of Saint John the Baptist, after sunset when the moon had first become visible, a marvellous phenomenon appeared to five or more men while sitting facing it. Now there was a bright new moon, and as usual the horns protruded to the east; and lo, suddenly, the upper horn split in two. From the middle of this division a firebrand burst forth, throwing over a considerable distance fire, hot coals and sparks. Meanwhile the body of the moon which was lower [than this] writhed as if troubled, and in the words of those who told this to me and who saw it with their own eyes, the moon throbbed as a beaten snake. It then returned to its former state.

This phenomenon was repeated twelve times and more, the flame assuming various twisting shapes at random then returning to normal. And after these vibrations it became semidark from horn to horn, that is, throughout its length. Those men who saw this with their own eyes reported these things to me who writes them; [they are] prepared to give their word or oath that they have added nothing false to the above.

Those last words underline the significance of this strange event. It is not being told second- or third-hand; it's not some wild account by a traveller returning from an exotic clime to report he saw a headless race of people with eyes in the middle of their shoulders. In this case the witnesses reported directly to the monk who was keeping a detailed chronicle of events in the monastery. Gervase himself is unremarkable. Apparently he never left the monastery of Christ Church, Canterbury, for any length of time; he wrote about church matters such as disputes between the archbishop and the monastery, but also recorded unusual events, everything from the burning of the cathedral in 1174 to a solar eclipse in 1178, and especially this strange eruption of the moon.

Gervase's account of a partial eclipse of the sun in September of 1178, just a few months after the moon incident, is reassuring to anyone who would doubt his ability to record astronomical events accurately. In that case he even points out how, as the moon passed in front of the sun but slightly off-centre, the sun appeared to have horns, first pointing towards the west, then down then to the east. This is exactly how that particular eclipse would have appeared from his vantage point.

Modern reconstructions of the night sky of June 18, 1178, in the Julian calendar, confirm that this fantastic event could have happened just as described. In Canterbury on that date the sun set about 8:15 local time, leaving behind a slim crescent moon just a day and a half old; that moon set forty-five minutes later. It was some time during that forty-five minutes that the "marvellous phenomenon" was seen.

If we take the account at face value, it needs to be explained. What could have happened to the moon on that night that appeared to engulf it with flames, then darkness, and made it look like a writhing snake?

In 1976 a scientist then at the State University of Stony Brook, Jack Hartung, published a startling explanation for this event. He proposed that our twelfth-century friends had seen the fireworks produced when a giant chunk of space rock crashed into the moon. He then made the leap to argue that the impact left behind a twenty-two-kilometre-wide crater which astronomers knew well and had named Giordano Bruno (after the renegade monk who was burned at the stake in 1600 because he had espoused the idea that the earth moved in orbit around the sun, not vice-versa). If Hartung is right, this is the only time in recorded history that such an impact has been seen from the earth. (Comet Shoemaker-Levy crashed into the planet Jupiter in 1994, but the fragments simply disappeared from sight into the Jovian atmosphere long before they would have hit a solid surface.)

Hartung was convinced the evidence was persuasive. He first interpreted the eyewitness description to fit his theory. "The upper horn split in two" implied that the upper part of the crescent moon (and this was a very thin crescent) was obscured either by the debris of an impact or the shadow of that debris. "Flames, coals and sparks . . ." were obviously the visible signs of a violent explosion, and "The moon throbbed as a beaten snake" meant that some sort of temporary—and turbulent—atmosphere was whirling around in front of the moon, disturbing its image in the same way that rising hot air makes objects behind it appear to shimmer.

So far so good, but the *Chronicle* continues, and, as Hartung admitted back in 1976, begins to stretch the interpretation. "The flame assuming various twisting shapes at random . . ." is problematical because the flame shouldn't last as long as the clouds of gas or dust that caused the moon to "writhe." Hartung isn't sure whether this so-called flame should be interpreted as a literal fire— incandescent material as he put it—or the flame-like glint of sunlight reflecting from vast clouds of dust particles. And finally, "it became semi-dark . . . throughout its length" is taken to mean that so much dust arose from the moon's surface that it blackened the entire visible surface.

And there you have the original event, as described in Gervase's *Chronicle,* interpreted as the crash landing of a large space object on the moon. But obviously that isn't enough. If there had been an impact of those dimensions, there should be a crater left behind, and that was the second part of Hartung's argument.

He began by trying to pin down the possible location of an impact. It happened at the "midpoint of the upper horn" of the extremely thin crescent of a new moon—in only its second day in the sky. Hartung took those words to mean that the "event" occurred at the extreme right edge of the crescent (and maybe just past it), about halfway from the lunar equator to the north pole. For comparison, if you held a globe of the earth in your hands and looked straight at the Americas, an equivalent of the 1178 impact would strike just east of Bordeaux, France.

That is the "where" of the crater. And what sort of crater would it likely be? Hartung estimated that the smallest cloud or shadow on the moon that could be seen by Gervase's friends would be about forty kilometres across. But obviously what the men saw was much more than just barely visible. So Hartung did some celestial mechanics and came up with the following: the explosion and subsequent cloud should be pegged at one hundred kilometres across; that in turn means there should be at least a ten-kilometre-wide crater. That is what's left behind after the cloud material is excavated by the explosive impact. Moreover that crater should exhibit signs of newness, most importantly a set of bright straight lines called "rays" radiating out from the crater.

One good look at a map of the moon revealed what they were looking for: a crater, Giordano Bruno, that had attracted attention years before because of its youthful appearance. It is twenty-two kilometres across and is the source of bright rays extending hundreds of—in some cases more than a thousand—kilometres across the lunar surface.

Its precise location is 36 degrees North and 103 degrees East, a little closer to the equator than might have been expected, but just around the right edge of the moon as Hartung had predicted. And the bright rays are an important piece of evidence. Rays are assumed to be material flung out in all directions from an explosive impact. They are bright because they haven't been weathered as much as the rock around them—the ray material was underground prior to the impact and so was sheltered from the solar wind and micrometeorite hits.

Giordano Bruno has an absolutely spectacular set of rays, longer for its size than any other crater on the moon. If rays become more indistinct with time as they weather, and so appear to shorten and darken, then Giordano's rays suggest a very recent impact. So this crater is in the right place and might even have formed at the right time.

But there are no other records of the event that have surfaced. And seeing any huge impact on the moon is extremely unlikely: the vast majority of hits from asteroids or comets occurred three to four billion years ago, when the young solar system was full of such stuff floating around. There are still killer asteroids around, but most of them long ago impacted something somewhere. Today they are thankfully rare.

So there it was: a fantastic story, but one accompanied by lingering doubts. In some scientists' minds, these were far more than doubts, they were the conviction that Hartung was just plain wrong. Two researchers at the American Meteorite Laboratory in Denver took issue with the story just a few months after his ideas were published. H.H. Nininger and Glenn Huss agreed that there was no reason to doubt the authenticity of the report, but they couldn't bring themselves to believe that the flames, sparks and smoke came from a lunar impact. They doubted that five men sitting together would even have noticed a hundred-kilometre-long cloud on the moon, and did their own calculations to show that the flames would have had to travel hundreds of kilometres from the site (on the dark side of the moon) where the impact occurred to a place where they'd be visible from earth. As they put it, "How else could 'fire, hot coals and sparks' large enough to be seen and *differentiated one from another* (italics theirs) from a distance of 376,000 km be formed and spewed out 'over a considerable distance'?"

They weren't finished with the fire and sparks. Nininger and Huss were also troubled by Hartung's idea of a temporary atmosphere, wondering not only how the amount of gas from such a cratering event could obscure eighteen thousand square kilometres to a depth visible from the earth, but also how such a smog-like atmosphere would be produced in the first place. Their concern was that most of the material involved in the blast would fall back very soon under the influence of the moon's gravity.

An impact seemed an unlikely explanation to these two. They opted for an explanation that Hartung himself had briefly considered, then rejected. Nininger and Huss claimed that what our twelfth-century witnesses saw was a meteor in our atmosphere, and that it just happened to be in the line of sight of the moon.

Remembering that the new crescent moon would be close to the horizon, Nininger and Huss envisioned a meteor that entered the earth's atmosphere travelling almost parallel to the earth's surface—that would allow it to remain visible for up to a minute. If it were moving more or less directly towards the witnesses, it would remain centred on the moon (although the two meteorite experts even suggested it might have been travelling in a tight spiral, thus accounting for the movement of the flames in apparently random motions).

Meteorites apparently can put on a spectacular show like this, even though most of them are the short-lived shooting stars you see occasionally in the night sky. Also, medieval people, living in an era when night skies were undiminished by city lights, were likely more familiar with shooting stars and meteor showers than most of us are today. The problem with the meteorite idea is that it is a one-in-a-million situation: a meteor just happens to be coming directly at witnesses and also just happens to be superimposed on a very skinny, short-lived crescent moon. The odds against this happening are pretty high.

If this story had amounted to just these two rival descriptions, it wouldn't have been much of a story. But unless some as-yet undiscovered medieval account of the same meteor in a different position *vis-à-vis* the moon is uncovered, Nininger and Huss's account can't be proved or disproved. On the other hand, there have been attempts to gather more data for or against Hartung's crater.

One in particular came out of left field.

Odile Calame and J. Derral Mulholland, who worked together in the late 1970s at the Centre d'Études et de Recherches Géodynamique et Astronomique in Grasse, France, published a paper in 1978 in the prestigious journal *Science,* backing Hartung up on two counts. First they produced evidence that convinced them the impact could have been seen easily from the earth. But much more dramatically, they suggested that the moon was still ringing like a bell from the impact.

In addressing the question of visibility they confronted Nininger and Huss head on. They tried to equate what was necessary for visibility to what could have happened, and satisfied themselves that a visible impact was well within the realm of possibility. Given that the rays extend hundreds of kilometres from the Giordano Bruno crater, Calame and Mulholland argued that a significant percentage of the total debris thrown up by the impact could have travelled that far, and that if the incoming chunk of rock had been travelling about twenty kilometres *per second*—not an unreasonable estimate— much of the debris could easily have traversed the visible crescent of the moon. Calame and Mulholland describe the event as not only being visible from Canterbury, but as "apocalyptic" in appearance.

However, their claim that the moon is still vibrating from the impact—today—was much more intriguing. For this argument they used data from laser ranging equipment left behind on the moon by the Apollo astronauts. Lasers aimed at those reflectors measure the distance from the earth to the moon with amazing accuracy. If the moon were still vibrating from a hypothesized Giordano Bruno impact, its surface would be moving very slightly but regularly back and forth. This is much more complicated to measure than I've made it sound, partly because there are at least three different ways the moon could wobble or vibrate, and two of those signals are impossible to resolve using the laser data. But one type of vibration, a regular change in the moon's rotation rate, can be detected with the lasers, and they reveal a pulsing, a back-and-forth movement of the moon's surface, a few metres in magnitude, consistent with what would be expected from an impact the size of Giordano Bruno eight hundred years ago.

Even the authors admit this is not a confirmation of Hartung's theory—it's more like the absence of a refutation, but in science that counts too. And the story doesn't end there.

In 1994 a little spacecraft called Clementine (because it was to be "lost and gone forever") took up an orbit around the moon and began to map the surface in eleven colours, some visible, some in the infrared. Doing so makes it possible to do remote geology, because the colour of the moon changes from place to place depending on the kind of surface material, and—more important in this case—on the amount of time it has been exposed to space.

Clementine produced fabulous images of the Giordano Bruno crater. They confirmed that it is very fresh. These images are the bluest of all the crater images gathered, significant because as moon materials are exposed to space and start to weather, they get darker and redder. (Rocks are weathered mostly by charged particles arriving in the solar wind and micrometeorite hits, each of which creates a miniature shock and heat wave.) Farther from the crater wall, fresh new material is apparently mixed with older, more weathered rock and soil, as you'd expect.

But there is one nagging problem. The Clementine images reveal some reddish, weathered material on the slopes of the wall of the crater. Most space geologists assume that visible signs of weathering shouldn't appear as early as eight hundred years after the material is exposed. On the other hand, no one really knows how rapid weathering is—they're just guessing. But either weathering is much more rapid than anyone thought, or Giordano Bruno is much older than eight hundred years. And at the moment there is no better resolution to the mystery than that. But with recently renewed American interest in visiting the moon, the possibility remains that detailed imaging from lunar orbit or even the collection of actual samples from Giordano Bruno could resolve the mystery.

Even if we find out what those five witnesses saw, we will certainly never know what they thought about the event. If they were educated people, they likely knew about the celestial spheres that encircled the earth and were made of ether, or the "fifth essence" the "quintessence," the pure, unchanging stuff in which the stars and planets are embedded. The moon was at the border between this astronomical perfection and the grubbier, ever-changing earth—its imperfections were clearly visible as the craters and dark areas that make the "man in the moon." So an explosion on the moon might not have shocked them as much as seeing a similar event on a distant planet.

Those less well-educated might have feared the event had they seen it; the moon was known to have powerful influences over the earth (the tides) and believed to be capable of the same with people's lives. It was a commonly held belief that the course of an illness could be influenced by the phase of the moon. It would have taken a serious optimist to interpret this cataclysmic event as anything other than malevolent.

Today our approach to this event is to try to reconstruct it, to imagine (guided by the science) what it would have been like. The impact of the asteroid that created Giordano Bruno would have been a spectacular event, whether it happened in 1178 or not. A chunk of rock a kilometre or more across, travelling at something like fifteen or twenty kilometres a second, would slam into the lunar surface, creating shock waves travelling both forward into the moon and backward into the asteroid. The asteroid would vanish instantly; but the solid rock immediately under it would be vaporized too. Rock that wasn't changed immediately into gas would melt under the intense pressures generated; the crater would be splashing and flowing. The gases, dust, liquid rock and solid rock would explode up and outward completely coating the immediate vicinity of the crater with debris; farther out (tens of kilometres in this case) the fallout would be intermittent; even farther out (a thousand kilometres) the rays would be laid down. It would be an event unlike any seen on earth, perhaps the same as a 100,000 megaton bomb—ten million times more powerful than the bombs that destroyed Hiroshima and Nagasaki. The witnesses in Canterbury may have been the last people on earth—and maybe the only people *ever*—to have witnessed such an explosion.

Natural Battles

Antlion King

Although I work on the Discovery Channel's daily science newsmagazine, @discovery.ca, I would be the first to admit that the images we use to promote the channel are not the experts we interview about the latest breaking news from the world of science, charming and entertaining though they might be. No, when you think of Discovery Channel, you think of sharks attacking sea lions, leopard seals prowling around in Antarctic waters waiting for unfortunate penguins, and alligators exploding from the water to down any wildebeest foolish enough to cross the river at the wrong time.

What exactly it is about the predatory moment that stirs us I don't know: perhaps we have some dim species memory of being prey ourselves (although some people I know are much more likely to identify with the predator). But these examples of life and death on the Serengeti or in the world's oceans ignore much of what is really interesting about the relationships between the hunter and the hunted, because we are captive to human scales of time and size.

The chase is only one final moment for two individuals; the habits, physique and sensory apparatus of the animals have been evolving together for millions of years either to avoid or create these few seconds of drama. Regardless of who triumphs this time around, there will be more chases, more ambushes, and as the total score accumulates, the destiny of the species is written. But not only do we distort the time scale of the struggle by concentrating on a single attack—we are oblivious to some of the most dramatic struggles between predator and prey by concentrating on those that are big, swift and strong. The drama of the struggle for life can be seen much more clearly on a smaller stage.

One of my favourites is the never-ending battle between ants and antlions. It will be a long time before Disney releases *The Antlion King*—these may be beasts, but they are a little too alien to play hero on the theatre circuit. Antlions are actually the larval stage of insects which resemble damselflies, which in turn look like dragonflies except they clasp their wings together over their backs when resting. The adult antlion, though it has those large wings, is a weak, bumbling flyer, most often seen fluttering around lights at night. It lives about three weeks and is an unremarkable creature.

Its larva is a different story. It looks not at all like the winged adult. It comes in many different shapes and sizes although it's usually not much bigger than its prey. In general it is a flat, stubby wingless creature, covered with an array of short spines, and armed with one outstanding physical feature: two vicious meathook-like mandibles extending out to either side of its head that clamp together like the pincers of a lobster. It is, like the rattlesnake, an ambush predator, but a unique one. The antlion finds a patch of sandy soil, digs itself a pit, then buries itself at the bottom of the pit with only part of its head and the fearsome mandibles showing. An unfortunate ant comes along, stumbles down the sloping walls of the pit right into those jaws, which snap together instantly. If the ant is one of the lucky ones to escape the initial attack, it will begin to scramble frantically back up the walls. The antlion responds by using its head to flick sand at the hapless ant, causing it to tumble back down.

When the capture is finally made, the antlion sucks the juices out of the ant, and in a final display of arrogance, tosses the dried-out shell of its victim out of the pit. Of course arrogance is attributing a little too much humanness to the pinhead antlion, but it's not just me. In 1732 the Abbé Pluche, in a book called *Spectacle de la nature*, describes the expulsion of the dead ant this way (in an English translation from the 1740s):

> When he has gratified his voracious appetite, and nothing more remains than the carcase, drain'd of all its juices, he takes particular care to remove and conceal it. The sight of a dead body might probably prevent some future visits, and render his habitation a place of ill repute; for which reason he extends the mangled carcase on his horns [the mandibles], and with a sudden jerk throws it about half a foot beyond the edges of the trench.

It all seems devilishly clever, but in this case predator—and prey—are operating with extremely tiny brains and therefore a need to run their lives on automatic pilot. They are tiny chitinous robots, not thinking their way through these dramatic events so much as acting. And that means that the scripts they're using— edited by millions of years of evolution—had better be good. Of course if they weren't, antlions wouldn't be around for us to observe. But even if their behaviour has none of the flexibility and innovation of the lion or the cheetah, it is nonetheless superbly adapted to its goal of, ahem, sucking on ants.

It all begins with the creation of the pit. Most of the time antlions make their pits in sand (although they have been known to use dust or even coal ashes) and although different species around the world go about this in different ways, the North American antlions in the genus *Myrmeleon* (including the one with my favourite name, *Myrmeleon immaculatus*) dig pits in pretty much the same way. These species are odd, even in the insect world, because they spend most of their time walking backwards. When an antlion is ready to make its pit, it starts backing through the sand, just under the surface, madly flicking grains of sand into the air with its head as it goes. As it moves backwards, it leaves a shallow furrow behind (actually in front of it). It looks as if you had just dragged your finger lightly through the sand. Sometimes these paths wander here and there, a habit that prompted the nickname "doodlebug."

Even this simple act is made easier by the shape of the antlion's body. It tapers from front to back and the narrower rear end is therefore able to slide through the sand easily. It also has tufts of hair on its legs and body that bend easily when the animal is moving backwards, but stiffen and anchor the body when it is bringing its legs back for another push. Even the arrangement of the tufts on the antlion's body is adapted to pit-digging: there are two rows of longer tufts on the underside and when it drags its body through the sand, these rows clear two parallel paths through which the animal can move its legs on the return stroke.

Backing up randomly through sand is a long way from creating a pit. To accomplish that the antlion has to change tactics, and it begins to move in a circle, using its abdomen to plough through the sand and its head as a shovel for flicking sand grains away. As it continues to circle, it spirals inwards, digging deeper and deeper, excavating the sand, until it finally has created a pit with sloping walls. The antlion then digs itself into the sand at the bottom of the pit, with only the mandibles protruding. And there it waits, sometimes for days, before its prey arrives.

While I have painted a picture of this creature being captive to its own inflexible behavioural routines, there can be many of these, allowing the beast to cope with a variety of troubling situations. For instance, if the antlion happens upon a small stone while it is digging its pit, it removes the stone. But sometimes that requires actions worthy of a gymnast-weightlifter. Again the eighteenth-century European naturalists led the way in recognizing the antlion's prowess. Charles de Bonnet (a man known better for the hallucinations he suffered later in life, now called Charles Bonnet syndrome) was the first to conduct the "Sisyphus" experiment. Sisyphus was the unfortunate mythical soul who was not only condemned to life in the underworld, but had to perform the eternal task of rolling a rock up a hill, only to have the rock fall back just as he reached the top. Charles de Bonnet placed a pebble in the antlion's path, and watched as the insect balanced the pebble on its back and carefully backed out of its pit:

> At any moment the burden is liable to fall off, either to the right or left, or even to roll down over the back of the insect. It is only by suitably raising or lowering certain parts of his segments that he succeeds in keeping the pebble poised on his back. Yet notwithstanding all his strength and notwithstanding all his dexterity as a juggler, the pebble sometimes escapes and rolls to the bottom of the pit.

Bonnet noted that however many failures the antlion endured, it always returned to the task, and finally succeeded in removing the pebble. An American scientist, Howard Topoff, pushed the bar a little higher in the 1970s by repeating the Sisyphus experiment using perfectly spherical—and therefore extremely hard to balance—ball bearings. Each antlion performed admirably even with these highly unnatural objects, pushing the bearing up the sides of its crater adjusting its abdomen quickly from one side to the other, reminding Topoff of the way we would balance a broomstick on the tip of a finger. Topoff had earlier noted that if an antlion encountered a similarly offensive object while in the process of excavating the pit, it would heave the object out, but use different techniques depending on the size of the object.

The antlion would get rid of smallish pebbles (ranging up to five times its own weight) by positioning its head directly underneath and flicking it backwards, up and over its body and out of the pit. However, this became more difficult if the pebble were heavier. Remember, if the antlion is still digging the pit, it is somewhere on the side of the slope, working its way down. Flicking a pebble directly back over its head may mean that pebble has to traverse the entire diameter of the pit. So when a heavier pebble appears (one that is as much as eight times bigger than the animal) the antlion throws the pebble sideways over the nearest edge of the crater. For even heavier stones the antlion rotates itself so that it can use the combination of its most powerful backwards flick and the shortest distance to the edge of the pit. These experiments do not demonstrate that the animal is thinking, but they do illustrate that even if behaviours are genetically determined, rigid and uncreative, if they are present in the right combination they equip an animal with the ability to cope with all but the rarest obstacles to survival.

In the antlion's case, there is much more to behaviour than these relatively simple examples of throwing techniques. This predator depends as much on physics as its own biology. For instance, if the antlion digs its pit in sand that includes a variety of grain sizes, the finished pit turns out to be lined with the finest of those grains. There is a simple reason for this: as the antlion backs around in its spiral pit-making mode, it creates a furrow. The heavier grains on the walls of the furrow fall to the bottom, leaving the furrow lined with fine-grained sand. Eventually the inward-spiralling furrow resolves itself into a pit lined with fine sand. Handily, the angle of the furrow walls is high enough that the heavier grains will separate themselves out in this way.

There is another factor operating to ensure that the walls of the pit are lined with fine sand. Jeffrey Lucas of Purdue University examined the sand grains that had been thrown out of the pit and found that, paradoxically, bigger grains had been thrown farther from the pit. The explanation for this lies in the physics of airborne particles as described by Stoke's Law. The drag (air resistance) on a flying object is higher—relative to that object's momentum—for a smaller object. The antlion throws particles of any size with enough velocity that they will sort themselves in the way Lucas found: the bigger the particle, the farther from the rim of the pit.

This is not just an incidental bit of scientific trivia. Lucas also found that ants take much longer to escape from a pit lined with fine sand than one with coarse particles. So when it comes to capturing prey, physics seems to be on the antlion's side, but, and this should come as no surprise, the antlion is not standing dumbly by. During construction of the pit the antlion throws sand grains out at an angle of about 45 degrees, an angle that will maximize the distance large grains travel. So far so good. But later, after it has captured an ant and sucked it dry, the antlion is faced with a different kind of problem. It has to eject the carcass and the excess sand that has gathered at the bottom of the pit during the struggle between predator and prey. So the antlion swings into action, throwing the husk of the ant and the biggest particles at the preferred 45-degree angle, then suddenly switching to a higher angle—about 60 degrees—ensuring that finer particles that have come to rest at the bottom of the pit will arc through the air and land again on the walls of the pit. In addition, the animal throws the stuff it wants to get rid of in a straight line, but when it's tossing fine sand back onto the walls, it swings back and forth, in effect spraying the sand back onto the pit walls.

There is even data suggesting that the antlions regulate the velocity of the particles they eject from the pit, tossing out sand particles relatively slowly during the construction of the pit, but accelerating them during cleaning. Why? During pit construction, the antlion is removing sand only, and if the smallest particles don't quite make it beyond the pit walls so much the better—they contribute to the pit. But during cleaning, unwanted particles should be heaved out with the greatest speed possible, so they land far away from the pit and are not likely to fall back in.

This subtle intertwining of behaviour and physics is captivating: the antlion may be the nearest thing to an automaton, but it is a very well-designed one. If you are not yet convinced by the picture I have presented so far, there's more. The same Jeffrey Lucas who measured the speeds and angles of objects being tossed out of the pit went on to examine the structural details of the pit itself. And while most descriptions of the antlion pit call it an inverted cone, it turns out (at least for some species) that the cone is not exactly symmetrical.

Lucas discovered that the front wall of the pit, that is that part of the wall facing the antlion, is steeper by about eight degrees than the wall behind the antlion. The walls to either side were in between, so the pit is a funnel, but tilted towards the antlion. A closer look revealed that the composition of the walls was different as well. The front and side walls, the steeper versions, were also covered with the finest sand, the shallower back wall incorporated some coarser grains. Remembering that ants that blunder into the pit have a more difficult time scrambling up walls composed of fine sand, and adding the reasonable idea that they would also have more difficulty with a steeper slope (mostly because it's easier to start landslides on them), the picture that is being painted here is of a pit that offers differing degrees of difficulty for escape. In particular, the front wall sounds impossible; the back wall, shallow and coarse-grained, sounds inviting.

So why is it then that ants seem not to take this into account when trying to escape from a pit? By carefully tracking the behaviour of ants which had eluded an antlion's grasp at least once, Lucas was able to show that while some ants just made a beeline (!) for the edge of a pit, the majority changed direction at least once, and that majority tended—surprisingly—to head for the front slope, the one that was the steepest and should be the most difficult to climb.

Seemingly senseless, but actually a good strategy for survival. If an ant escapes its initial brush with the jaws of the antlion and starts to scramble back up the walls, the antlion will start madly flinging grains of sand at the ant in an attempt to bring it back down. But antlions are best at digging their heads into the sand and pitching that sand over their backs, i.e., towards the back wall of the pit, the one whose shallow coarse-grained expanse is so inviting. Ants which choose that route are usually making a bad mistake, and apparently genes for such unwise behaviour were long ago weeded out of ant populations. Today's ants avoid the siren song of the back wall, and try desperately to climb the front. They won't get showered with sand there (and in the unlikely case that the antlion actually decides to give chase, they will be safe for a few moments on that slope, because the antlion can only move *backwards*), but the steep slope and its fine grains might prove their undoing anyway. That the ants are all too aware of the presence of the antlion was shown by creating pits with no antlion present; ants stumbling into those didn't hesitate to take the shallow back-wall route to safety.

Even the actual capture and kill, when examined closely, reveal as yet unexplained tricks on the antlion's part. The capture is a gruesome piece of business: if an ant falls to the bottom of the pit and is grasped in the antlion's jaws, some considerable time may pass before the lion actually begins to suck the ant dry. In some species the antlion locks one jaw into place, holding the ant tightly, then searches with the other for an opening between the shell-like plates covering the ant's abdomen. If the probing point of the jaw finds such an opening, it plunges in, and begins to inject poison. It's a give-and-take situation: give some poison and digestive enzymes and take some nutrients, those nutrients unfortunately being the contents of the ant's body.

Even at this desperate stage the ant has one last card to play. Ants which are members of the subfamily *Formicinae* are able to spray corrosive formic acid at their enemies. Antlions are extremely sensitive to the acid: in experiments conducted by Tom Eisner of Cornell University, antlions which had a single drop of formic acid (the amount they'd get from an ant) dropped on their heads from a micropipette released their ant prey within three seconds. Nonetheless, antlions which captured these acid-throwing ants never once released them—in each case they killed and sucked them dry.

On closer examination of the dried-out corpses, Eisner and his colleagues discovered that the antlions had managed to suck the entire contents of the ant dry (if they had had bones they would have been accurately described as being nothing but skin and bones; as it was, they were virtually nothing but exoskeleton). However, at the same time the antlions had avoided puncturing the acid sac, a delicate feat considering that the food crop and the acid sac are both thin-walled fluid-filled bags.

So antlions are good at avoiding the acid sac, but why didn't the ants spray it on them before they died? Here Eisner noted that in most cases when such ants attack humans, they bite first, then curl their abdomens around and spray the acid into the wound. Most of the time the ants must bite first or they simply won't spray. Exactly how the antlion prevents that key bite isn't known— maybe considering most of the antlion is buried, it is difficult for the ant to know just where to bite.

If this had been antlion versus ant as a Discovery Channel television documentary, it would have been almost entirely shot in close-up. And that might give the impression that the overwhelming advantage lies in this case with the predator. But now it's time for the final scene, the equivalent of the sun setting as thousands of wildebeest make their way across the Serengeti. As the camera pulls back from the antlion pit, other pits come into view, then others, more and more antlion pits. This is typical— where the environment is suitable, the *Myrmeleon* larvae dig. But where are the ants? Yes, it turns out the ants do have a trick up their sleeve. Antlions are ambush predators: they are out of luck if the prey just doesn't bother visiting the neighbourhood. There is good experimental evidence showing that where the pits cluster together, the ants take a pass. This prompts one final interesting question.

Ants communicate very effectively using chemicals called pheromones. A foraging ant finds a source of food, and lays down a chemical trail on its successful return. Other ants follow the trail outward, find the same food cache and lay down their own trails. The ant that arrives to find the food source consumed does not lay down a trail, and no other foragers are lured to a now uninteresting site. But how do ants, if indeed they use chemicals, inform their sisterhood that one of their usual foraging areas is now a mine field of antlion pits? What is the syntax of that message? It is established that they can do this, but we are a long way from understanding how.

And so the see-saw relationship between this predator and its prey is maintained. The closer an ant gets to the antlion, the worse its predicament—clever engineering and good biology ensure that. But an ant that doesn't show up won't get caught.

The Bacteria Eaters

So, naturalists observe, a flea
Hath smaller fleas that on him prey;
And these have smaller yet to bite 'em,
And so proceed *ad infinitum*.

These four lines from Jonathan Swift's "On Poetry: A Rhap-sody" reveal that while Swift might be best known for his understanding of parasitic relationships within his own species (the next two rarely quoted lines are: "Thus every poet in his kind/ Is bit by him that comes behind;") he was well aware of the pyramid of life in the wider biological world.

Swift's readers would have appreciated his reference to fleas. At a time when the citizenry wasn't exactly pushing the envelope of personal hygiene, all were familiar with the concept of a tiny crea-ture preying on a much larger host. They would have been at least faintly amused by the idea that the fleas that tormented them might likewise be suffering. But neither Swift nor his readers could have imagined just how appropriate was his use of the words *ad infinitum.*

It required Swift's imagination to conjure up the series of fleas biting fleas. The wonders of the microscopical world had already begun to be revealed by the time he wrote these lines, but just barely. The microscopes of the time were relatively crude optically but more important, because they depended on visible light for illumination, they were limited by the finite size of the wavelengths of that light—anything shorter than a light wave is invisible.

As we now know, Swift's infinite series of fleas continues into ultramicroscopic realms where even the most powerful instruments available give only limited glimpses of the struggle for life and where much of the evidence for the existence of strange life-forms is indirect.

My favourite inhabitant of that world is a virus, but not one that preys on human beings. They too are marvellous, but the virus that first captured my imagination—and still holds it—was something called a bacteriophage, a "bacteria-eater." Now simply called phage (and pronounced either *fawzsh* or *fayj* depending on whether you want to preserve the original French), the existence of these specialized parasites was first deduced early in the twentieth century. They were not even seen; their presence was inferred.

One of the discoverers of these beasts was a Montrealer by birth, Felix d'Herelle. When in Mexico in 1910 investigating a plague of locusts, he discovered that many of the insects were dying of a bacterial infection. When he grew cultures of these bacteria on plates of jellied nutrient in the lab, he noticed clear spots in an otherwise dense "lawn" of bacteria. Something was killing the bacteria in those spots, and that something passed right through filters that held back even the smallest bacteria. A few years later, after discovering the same phenomenon among the bacteria that cause dysentery in humans, d'Herelle concluded that an "invisible microbe" was responsible: a bacterial virus.

One of his first thoughts was that these viruses could be employed as a weapon against bacterial diseases, an idea that Sinclair Lewis explored in his Pulitzer Prize-winning novel *Arrowsmith*. Antibiotics turned out to be much more effective and the idea of phage therapy was set aside (although today as the problem of antibiotic-resistant bacteria becomes critical, interest in using viruses against them is heating up once again). However the "bacteria-eaters" held great fascination for biologists nonetheless because their simplicity offered a unique opportunity to unlock some of the fundamental secrets of life. That promise has been partly realized: phages have revealed much of the molecular underpinnings of life, although there is still much remaining to be discovered. What they have illustrated most vividly is that even in a world of molecular scale, predator and prey play familiar roles: one moves, the other countermoves.

One of the intriguing paradoxes of these beasts is that much of what we know about them has been determined from biochemical experiments, not by visualizing them. Felix d'Herelle, the pioneer of the research, could only have seen a bacteriophage near the end of his life—the first photos using the electron microscope were taken in the early 1940s. The electron microscope evaded the limit imposed by the wave-length of light by using a beam of electrons instead, giving this microscope the capacity to see objects a hundred times smaller than had ever been seen before. Even so, it has limits: specimens must be dried and prepared for viewing, so any phage seen in the electron microscope is dead, dried and long past being able to infect any bacterial host. Nonetheless seeing a phage is a revelatory experience, not only confirming the portrait painted by the biochemistry (the criminal suspect turns out to look just like the artist's composite drawing) but also reinforcing the idea that nature is endlessly inventive—and savage.

One more tricky issue before we enter this world: I've called it a "beast" but bacteriophage and all other viruses are not really alive. They can reproduce, but only inside a host cell, where they divert the cellular machinery from its normal tasks to the manufacture of viruses. Once outside that cell they are essentially inert chemical packages. They have great potential, but are incapable of realizing it on their own. They are on the borderline of life.

There are many bacteriophages, one or more for every kind of bacterium. They have been studied, not so much because they are interesting in and of themselves, but because they are relatively simple objects that can shed light on how genes work. The one that is probably the most intensively studied in a virus called T4 that parasitizes *E. coli*, the bacterium with the misfortune of being known mostly for its association with human feces—water quality tests search for the presence of coliform bacteria as an index of exposure of that water to human waste.

Indeed *E. coli* make up something like a quarter of the mass of all human feces, which given the fact that five thousand *E. coli* could be lined up across your little fingernail gives you an idea of just how many of these bacteria are in your gut right now, each one an inviting target for T4. It has been estimated that a trillion bacteriophages pass through the sewage plant serving a small city (60,000 people) every day.

Under ideal laboratory conditions an *E. coli* cell can divide every twenty minutes. Obviously, as has been pointed out many times before, that can't possibly be happening in the natural habitat (your intestine) or the earth would be swamped by these bacteria in a couple of days. Nonetheless coliform bacteria represent a highly evolved, incredibly efficient life form; thus any organism that would target it must be highly evolved as well, and

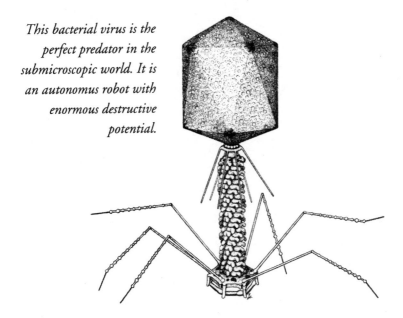

This bacterial virus is the perfect predator in the submicroscopic world. It is an autonomus robot with enormous destructive potential.

the T4 bacteriophage fits the bill. In fact it is speculated that T4 probably appeared on the planet shortly after its bacterial hosts, which puts its arrival at something like three and a half billion years ago. Its *modus operandi* substantiates the view that it is anything but primitive.

The unwitting victim, the *E. coli* cell, may be just visible at the limits of the ordinary light microscope, but it dwarfs T4, its killer. If an *E. coli* cell were a breakfast sausage, a T4 phage would be about the size of a black peppercorn. And remember, there could be five thousand such sausages laid end-to-end across your fingernail.

It's very easy to demonstrate in the lab just how devastating a T4 infection can be for *E. coli*. Bacteriologists can grow the bacteria on agar gel in a petri plate. If the gel incorporates the right nutrients, a population of *E. coli* will grow at remarkable speed, soon covering the surface of the gel completely with a lawn of bacterial cells. If one hundred phages were introduced to that lawn, you would soon see one hundred circular clearings (just as Felix d'Herelle did), areas where billions of bacterial cells have died, killed by one virus and its offspring. It's true that the bacteria reproduce quickly, but the phage does better: an original single virus infecting one bacterial cell will produce two hundred new viruses in half an hour. Each of those two hundred move on to infect another cell, and so it goes. That's impressive, but it is just numbers. The beast itself, the chase and the kill are remarkable.

A T4 phage looks a little like the Apollo lunar lander. It has a geometric head, a tail, and a set of tail fibres that spread out and attach to the surface of the bacterium. In function, however, it is more like a completely self-contained robotic spacecraft—fully preprogrammed.

The manufactured appearance—the unlifelike symmetry—is surprising at first look, probably because we think of microscopic infectors as tiny worms, or even miasmic gases, concepts left over from centuries ago. But the forces that dominate this world (where objects are millionths of a metre in size or less) are powerful short-range chemical bonds, and structures are nakedly molecular. For instance, a molecule will attract or repel others depending on the haze of electric charge surrounding its projections or the shape and orientation of tiny crevasses on its surface. A second molecule might fit like a hand in a glove or it might never make contact. This isn't to say that life in our world isn't dictated by the same kind of chemistry—it is. But other forces, especially gravity, play a dominant role. In the world of the phage, chemistry is it. In the description that follows, that's worth keeping in mind.

In the absence of prey, the T4 phage simply drifts with the tide—it is not capable of seeking out *E. coli*. In drift mode the tail fibres are stowed, pinned up alongside the tail. However, when the virus comes into contact with the surface of the bacterial cell, the tail fibres immediately swing down and spread out, and are the first parts of T4 actually to touch the *E. coli* cell. They will attach wherever they contact a specific receptor molecule that's part of the external coat of the bacterium. However, the bond between one tail fibre and its receptor is weak, too weak to anchor the virus. There are six such fibres and at least three must make contact before capture is complete. That doesn't happen immediately because the receptors are distributed across the surface of the bacterium like occasional repeating tiles in a mosaic.

This is the first step of what phage scientists call the "phage mating dance." T4 walks across the surface of its intended victim, tail fibres attaching, then detaching, until finally it makes sufficient, and permanent, contact. If *E. coli* had a brain, surely the unearthly sensation of phage tail fibres brushing lightly against its surface would inspire terror.

Once anchored, a remarkable series of events ensues. The virus adjusts its position so that the tail is positioned over a thin portion of the surface of the bacterium. Tail fibres attached to the flat base plate of the tail extend and pin the virus down (no escape now) and suddenly the base plate itself mysteriously changes shape from hexagonal to star-shaped. This triggers a rearrangement of the molecules of the outer sheath of the tail; the sheath contracts, the tail fibres bend and the virus is pulled down closer to the cell surface. The core of the tail actually penetrates partway through the multilayered outer envelope of the bacterium, an event likely made easier by enzymes in the base plate that chop up some of the surface molecules in that envelope.

Now the head of the virus sits just above the cell. The head is a rigid hollow case in the shape of an icosahedron, a regular twenty-sided geometric figure. It contains the genes of the phage, more than one hundred and fifty of them, all linked together in one long thread of DNA. Long of course is a relative term, but the phage DNA, stretched out, would measure several hundred times the dimensions of the head. No one is yet sure exactly how that much DNA is packed into that tiny space, a space made tinier by the fact that special packing molecules are stuffed in there as well. But at this point in the phage mating dance, the DNA isn't going to be locked inside the head much longer.

When the hexagonal base plate changed its shape, it opened up a channel wide enough for a single DNA double helix to pass through. Now the huge string of phage DNA, its entire genome, snakes its way through the tail, through the bacterial surface envelopes, the rigid cell wall and into the interior. It's all over in less than a minute, this process that some researchers have likened to throwing a potful of spaghetti—one enormous strand—into a colander and having the end of that strand find its way through a hole and then feed itself through completely. The energy to do that has to come from somewhere, but it's not yet clear where. One thing is certain: once the phage DNA has entered the *E. coli* cell the poor bacterium is not long for this world. And it is about to suffer the indignity of contributing through its death to the multiplication of the phage.

It's simple really. Among the hundred-and-fifty-plus genes in the phage DNA are those that direct (through the molecules they make) the shutdown of almost all *E. coli* activities. However, the cellular machinery formerly used to make *E. coli* membranes, enzymes, structural protein molecules—the machinery that maintained the bacterium's pulse of life—remains unscathed and is instantly converted to creating new phages. The now commandeered bacterial cell becomes a factory floor for phage parts. As the minutes tick by, scaffolds for building new heads appear here, tail fibres there, baseplates over here. It might appear simple, but in fact some of these parts are composed of several different kinds of molecules. David Coombs, a phage biologist at the University of

New Brunswick, has called the base plate alone "one of the most challenging biological structures ever studied in molecular detail." Some phage parts spontaneously self-assemble from their components, but others must be engineered together under the guidance of yet more molecules.

A hint of the subtlety of engineering involved can be seen in the manufacture of new phage DNA. Naturally it's assembled using the machinery that *E. coli* used to make its own DNA. But what is it made out of? Pieces of *E. coli* DNA that were disorganized, then dismembered, mere minutes after the phage gained access to the interior of the cell. The phage manages to scavenge about twenty viruses' worth of DNA from host DNA. But the phage DNA is different in one important respect: one of the four DNA sub-units is decorated with small molecules that identify it as uniquely phage. It's suspected this protects the phage DNA from enzymes inside the cell that normally attack and destroy any pieces of foreign DNA that they happen upon. It may even protect the intruder's DNA from its own DNA-destroying chemicals. Because such recognition is a molecular touch-and-feel sort of process, DNA with these unusual decorations escapes.

Assembly continues in an ordered but rapid fashion. Fully mature heads are built around head scaffolds (which are then discarded), then stuffed with a complete set of genes. Tail fibres bond to base plates, tail cores to sheaths, base plates to tails, and before the half hour is out hundreds of new phages are ready to be released. One final enzyme is manufactured which chews away the bacterial envelope from the inside and the progeny viruses escape to begin the routine all over again.

How do any *E. coli* survive in the face of such diabolical evolutionary design? They might come up with alterations to the receptors that the tail fibres recognize, which would literally make them "invisible" to the phage, but there's good evidence that the phages can simply respond by altering their tail fibres to make them visible again. *E. coli* also makes a variety of defensive DNA-destroying enzymes, but T4 can evade many of those by decorating its own DNA, although there's likely an ongoing battle here, with *E. coli* cells swapping defence genes among themselves.

Perhaps the most effective defences are what are called "guests" hiding in the *E. coli* DNA. These are genes left behind in the *E. coli* chromosome by other phages or in some case by some unknown visitor. These alien genes will not permit the T4 to reproduce inside the *E. coli* cell, but this act of defiance is a noble one for the bacterium, because the bacterium dies in the process, reminiscent of the infamous phrase from the Vietnam War, "We had to destroy the village to save it." In this molecular version, however, death of the bacterium does insure that no new viruses will be produced from it.

That is the story of T4 to date, and it's worth remembering that this story has not been put together by spending hours at the microscope watching it happen—that cannot be done. Our knowledge of this wickedly intricate attack comes from generations of clever experiments, building one on the other. The best of such experiments come to be known in the trade by the surnames of the experimenters. A perfect example is one of the earliest: the Hershey and Chase experiment of 1952.

At a time when hardly anything was known about the events I've just described, Arthur Hershey and Martha Chase were trying to determine what ensues after the phage attaches to its target bacterium. So they grew *E. coli* in culture fluid containing radioactive phosphorus and sulphur, two key elements in the most important substances in any cell, DNA and proteins respectively. They then infected those bacteria with phage, which proceeded to incorporate the same radioactive elements.

Now came the crucial part of the Hershey and Chase experiment. They mixed their radioactive viruses with fresh—nonradioactive—

bacteria, allowed them time to attach, then put the mixture into a kitchen blender. The forces generated by the blender would likely tear any attached phage away from their bacteria. Following blending the bacteria were poured out onto petri plates and allowed to form a bacterial lawn. As usual, clearings appeared wherever there were active phage.

When Hershey and Chase searched out the radioactive elements, they found that most of the sulphur was still back in the blender, while most of the phosphorus had apparently stayed with the bacteria that had been plated out. It was already known that almost all the radioactive sulphur would be incorporated into protein, while the phosphorus went straight into making DNA. So their experiment demonstrated that the virus DNA had likely entered the bacteria (and so hadn't been separated from them by blending) but the protein (the head, tail, baseplate, fibres and so on) had remained outside and had been sheared off. They were able to conclude that DNA enters the cell but the rest of the virus remains outside, one of the landmark discoveries in the study of bacteriophage.

T4 and its kin may represent the end of the line for Jonathan Swift's line of fleas extending *ad infinitum*. There are phages that are much less complex than T4 —some containing at most a handful of genes—but to date there has been nothing found to prey on the viruses that prey on bacteria. But who knows? It isn't impossible that some tiny fragment of naked DNA, shorn of all the machinery for infection that T4 boasts, might be insinuating itself into the phage DNA and being copied as new viruses are manufactured—a virus-within-a-virus. Given the history of discovery, the fact that such an entity hasn't yet been found is almost a guarantee that it exists.

An Uneasy Bargain

This is the story of a gene, one of the eighty thousand or more in every human cell. This gene has travelled the world in the bodies of its unsuspecting carriers; it has been, and still can be, a child killer. But this gene long ago offered us a Faustian bargain, one that apparently we couldn't refuse.

This gene is a mutant, and it causes the disease originally called cystic fibrosis of the pancreas. Now just cystic fibrosis (CF), it is the most common fatal inherited disease of Caucasians. The outlook today for a CF patient is much better than it was in the 1960s or even '70s; now patients routinely live into their twenties and thirties and there's no reason why that upper limit won't continue to rise.

The CF gene is recessive, meaning its deleterious effects are masked in the presence of the normal version of the gene. Because your body cells contain two copies of every gene, even one normal version is enough to override the effects of the mutant. You can survive quite nicely with a mutant CF gene in every cell in your body, as the people called "carriers" do. Roughly one in every twenty-five adults of European background carries a single copy of the mutant CF gene; one in every twenty-five hundred children of the same background inherits the double dose of the gene—and with it, the disease.

The normal version of this gene directs construction of one of the many channels that penetrate the membrane surrounding every body cell, permitting the movement of molecular traffic. This particular channel is specifically designed to allow the movement of salt (specifically the chloride part of the sodium chloride molecule) in and out of the cell. The channel is assembled in the interior of the cell then shipped out to the membrane and plugged into place, like a tap in a keg. However, this channel is no tap; the movement of chloride in or out of the cell must be controlled with exquisite precision; there are times when it must be shut down completely. Accordingly the channel is equipped with sensors which are activated by messenger chemicals inside the cell. At least two of these sensors power the channel up (it takes energy to move substances through the cell membrane) and another is what scientists call a "regulatory domain"; when it is activated, it turns on the flow of chloride.

You cannot see this channel, at least in the normal sense of the word—it is just too small. Even the most powerful microscopes would at best show it as a darkish lump. However, if you put X-ray imagery together with what's known chemically of the channel, a much more precise picture emerges. From a distance the channel would appear to be a smooth bell-shaped structure, with the top of the bell sticking through the cell membrane to the outside of the cell. But a closer view would reveal that the smooth surfaces of the bell are actually lumpy, or more accurately, ropy. In fact they are not surfaces at all. The entire molecule is one unbroken chain, looped, twisted, spiralled and stuck together to create crevices, canyons and a variety of weirdly shaped extensions, all on a molecular scale.

The chain from which the chloride channel molecule is constructed is a string of 1,700 amino acids chemically bonded one to the next; the most common mutation creating cystic fibrosis eliminates just one of those 1,700, the amino acid that normally occupies position number 508. The removal of that one amino acid, one molecule out of 1,700 or the "deletion at 508," is catastrophic: it arrests the manufacture of the channel, leaving it in an immature, incompletely folded form, and the cell simply abandons it where it was made rather than transporting it to the cell membrane where it belongs. The flawed channel never functions.

Therefore the organs and tissues of those individuals with two such mutant genes contain cells with no chloride channels on their surfaces; carriers—people with just one mutant gene—have about half the normal number of channels on their cells. The difference between one and two such genes turns out to be crucial.

Even though a child with CF has no cells anywhere in the body with functioning chloride channels, certain organs are hit harder than others, primarily the lungs. Normally lungs are lined with a thin layer of mucus which, having trapped bacteria, fungi, dust and dirt is gradually expelled from the lungs by the beating of cilia that line the airways. In CF, that thin layer of mucus becomes thick and difficult to move (the exact link between mucus and defective salt transport is still unclear) and instead of acting as a conveyor belt for the removal of bacteria, the mucus provides a medium for their growth. Some of the bacteria that are the most common colonizers of the lungs of cystic fibrosis patients actually send chemical messages to the lung cells to accelerate mucus production. There is also evidence that an antibiotic normally active on the surface of the airways becomes inactivated by the high concentrations of salt, salt that wouldn't be there if the chloride channels were working properly.

The bacteria take up residence and cause repeated lung infections; the resulting inflammation, combined with the thick layer of mucus, causes the airways to narrow. The lung becomes the major focus for treatment: antibiotics for the bacteria, chest pounding to loosen the mucus. If lung infections become chronic, they can gradually destroy the lung tissues, causing respiratory failure and death.

The pancreas is also hit hard enough in this disease that some patients must take capsules of replacement digestive enzymes when they eat. The mutant gene even makes its presence known in a kiss. "Woe to the child which when kissed on the forehead tastes salty. He is bewitched and soon must die" is a medieval European saying that recognizes that excess saltiness in the sweat, a direct result of the inability of sweat-secreting glands to reabsorb salt just before secretion, is a deadly symptom.

There is one very puzzling aspect to all this. Until recently cystic fibrosis usually brought death before the age of twenty, and diminished health prior to that age usually precluded having children. (In fact something like 95 per cent of CF males are infertile because the fine-calibre ducts that carry sperm are blocked; female CF patients are sometimes infertile, too, because the entrance to the uterus is blocked by a plug of mucus.) This lack of reproduction means that almost always the two mutant genes in every CF patient had reached a dead end; every death eliminated two copies of the gene, and gradually but inexorably the disease should have died out. But nothing like that has happened: today 4 or 5 per cent of Caucasian adults carry the gene. It is much more frequent than it should be, and after trying out all the potential genetic explanations, scientists have agreed that paradoxically, this mutant gene, under some circumstances, must actually be beneficial.

There is one well-established precedent for this paradox. Sickle-cell anemia is caused in the same way as CF, by a double dose of a mutant gene. In the case of sickle-cell, which affects predominately people of African heritage, the mutant gene alters the shape of one of the chains in the hemoglobin molecule, causing separate hemoglobin molecules inside red blood cells to stick together in needle-like strands, in turn deforming the normally ovoid red blood cells into sickle shapes. These abnormal blood cells can become stuck in blood vessels and clog them, stopping the flow of blood and creating a crisis for the patient. However, people can live with one copy of the sickle-cell gene (meaning half their hemoglobin molecules are normal), although it makes oxygen-demanding activities like mountain climbing difficult. But the question in this case was the same: how could such a disadvantageous gene maintain such a high frequency in tropical Africa?

The answer was malaria. Having one copy of the sickle-cell gene made a significant number of that person's blood cells resistant to the malarial parasite (apparently because the sickling process is accompanied by chemical changes that kill the parasite). Having no sickle-cell genes would mean that while neither you nor your children would get the disease, you were susceptible to malaria. Having one copy meant you wouldn't die of malaria, and the chances were, depending on whom you married, many of your children wouldn't either. But if you married another carrier, each of your children would have a 25 per cent chance of getting sickle-cell disease. The trade-off was apparently worth it and the sickle-cell gene persists in parts of Africa where malaria remains a threat. A gene causing a similar disease, thalassaemia, is also malaria-protective in countries around the Mediterranean.

So with sickle-cell anemia as a model, all that remains is to identify the disease against which the CF gene is protective. But thirty years of guessing and experimenting have still not succeeded in doing so, although some remarkable details of the history of this gene have been uncovered.

One of the best candidates right now for the disease against which CF offers protection is cholera or a similar diarrheal disease, but while the molecular reasoning backing up this choice is a good fit, the history and demography of the disease are not. Cholera is, at the molecular level, a mirror-image of cystic fibrosis. While in CF there is no chloride flow (because the channels are absent), the toxin from the cholera bacterium, *Vibrio cholerae*, causes chloride-secreting cells in the intestine to open up and stay open. Salt and water come pouring out of the cells uncontrollably, causing a dangerously dehydrating diarrhea. The cholera toxin acts through one of the same molecular messengers that signals the normal CF-controlled channel to open. Paul Quinton, a CF expert in California, first suggested that cholera was the mystery disease in 1982. Then in 1994 a research group at the University of North Carolina bred mice that seemed to back up Quinton's idea. Mice with two normal copies of the CF gene responded as expected to cholera toxin by secreting copious amounts of chloride and water into their intestines; mice with one mutant copy of the gene secreted about half as much, and those carrying a double dose of the mutant gene (and so lacking chloride channels) produced no more than they would normally. It was a beautiful match of theory to data: the CF gene should protect against fluid and salt loss, and, at least in mice, it does just that.

But that experiment, while it has swayed many people, hasn't sealed the case for cholera. There is one major objection: CF is most common among people of European background, yet the areas of the world where cholera is the biggest threat are Africa and Asia; there, the mutant gene is rare. Remember the sickle-cell model: for a mutant gene to persist, it must protect against some disease that could attack the very people carrying the gene.

The Barmaid's Brain

It is also possible that the reason the CF gene isn't common in those countries (mostly tropical) where cholera is endemic could be that even carrying a single mutant gene reduces the amount a person sweats; losing that degree of cooling might be a greater disadvantage than the risk of cholera. But this is speculation only.

On the other hand, maybe cholera used to be a killing force in Europe, deadly enough to make carrying the CF gene advantageous; if that were true, while the disease is no longer common, its decline would have been so recent that we wouldn't have yet seen a subsequent decline in the CF gene. That is possible; cholera was likely one of the great killers of children in the Middle Ages together with plague, smallpox and tuberculosis. But the distribution of the disease before that is controversial. It used to be thought that prior to 1817 (the beginning of the first cholera pandemic) it existed only in India and Sri Lanka, but now it seems as if there are reliable records of cholera in Europe dating back to the 1500s and 1600s (and there are unverifiable hints of the disease even in ancient Greece). The last major epidemic in Europe was in 1866, when hundred of thousands were killed, and it continued to flare up in various European countries as recently as the 1890s. While cholera has faded now, it was still common too recently to have seen a significant decline in the CF gene as a result.

But even if cholera existed in Europe five hundred years ago, that isn't enough. There is genetic evidence pointing towards a much greater ancestry for the gene. The deletion at position 508 is by far the most common mutation underlying cystic fibrosis—something like 70 per cent of all northern Europeans carriers have it. But closer to the Mediterranean, other mutations are more common and the deletion at 508 accounts for no more than half the mutants. (Remember that the channel molecule is a chain of 1,700 amino acids; many of those, if changed by mutation, could render the entire molecule useless.) There are more than five hundred different mutations known for this single protein molecule, and this incredible variety opens up the possibility of mapping the historical movements of the gene.

Take the G542X mutation, a focus of damage some distance along the molecular chain from the deletion at 508. It comprises barely 1 per cent of all CF mutations in England and less than 0.5 per cent in Sweden. But further south it comes into its own. Eight per cent of all mutations in Spain are G542X, and if samples are taken just from the Spanish Mediterranean coast, that percentage soars to 14. G542X accounts for about the same percentage in Algeria but an amazing 21 per cent in Tunisia. Directly north of Tunis on the island of Sicily the frequency is 10 per cent.

Genetic analysis, combined with historical records, points to the Phoenicians as the people who brought the G542X cystic fibrosis mutation to the western Mediterranean, probably about three thousand years ago. Sicily, southern France and Mediterranean Spain all were occupied by these legendary sailors and traders; the city of Carthage (now Tunis) was a major Phoenician settlement. If these claims of Phoenician mutations are true—and they look good—then mutant CF genes like this one have been around for millennia.

In fact, tens of millennia might be a better guess. A group of geneticists from across Europe has studied the most common mutation—the deletion at 508—intensively; by tracking minor associated mutations and how they have changed over time, they have been able to estimate when the 508 mutation first appeared in Europe. They estimate the minimum to be 52,000 years ago (based on 2,600 generations at twenty years each) and suggest that the actual date is likely to be much older. What's more, the population which introduced the gene was itself of a different genetic background than modern Europeans.

Apparently the mutant gene arrived roughly when physically

modern people arrived in Europe in a big way, about fifty thousand years ago. These were the people who supplanted the Neanderthals in Europe. Some time after that, these original arrivals were in turn replaced by other, genetically distinct groups. Those later arrivals account for the fact that the 508 deletion is much more common in northern Europe than around the Mediterranean—in the south 508 was partially replaced by other mutations arriving with newer immigrants.

If the most common CF mutant gene is more than fifty thousand years old, then it presumably has offered its mysterious advantage for a similar length of time, a fact that unfortunately does not do much to solidify cholera's position as the best candidate.

So maybe cholera, despite the lab evidence supporting it, isn't the right candidate after all. The experiments on mouse intestine are persuasive, but not proof. There have been several other diseases considered, including influenza, typhus and tuberculosis. All have been important diseases in Europe throughout the centuries, but the exact cellular events involved in those diseases that could be countered by a mutant CF gene have not been well worked out. Nonetheless a quick look at a selection of the other candidate diseases reveals just how tricky this business can be.

Two of them are particularly intriguing for their unexpectedness. One is syphilis, the other asthma; together they exemplify the different approaches to solving this problem. Syphilis was practically an epidemic disease in Europe in the sixteenth century and as such coincides precisely with the geography of the mutant CF gene today. But this seems to be the only convincing link between the two. The author of this idea, microbiologist David Hollander, argued in 1982 that venereally transmitted syphilis (which can be acquired by a child in the womb) was a disease of temperate climate; that was the reason it flourished in Europe, and that in turn made it the best candidate for the CF effect. He couldn't have known that, years later, research would show that the CF mutant gene came to Europe, not centuries before the 1500s, but millennia. Hollander would now have to show that venereal syphilis existed among those pioneering European populations, who emigrated from warmer, perhaps even tropical climes. But the evidence for that isn't good. Many still believe that syphilis originated in the Americas and paleopathologist Bruce Rothschild has found what look like syphilitic lesions on skeletons as much as 1600 years old from New Mexico. On the other hand, evidence for syphilis in Europe prior to Columbus's return from America is scant. There are some skeletons from English graves that show the scarring of syphilis, but they might not predate Columbus. At the same time, experts can't agree whether some ancient Greek remains—which date back to hundreds of years BC—show certain signs of syphilis.

The suggestion that the CF gene might offer protection against asthma is a new and surprising idea. After all, both diseases cause breathing difficulties by narrowing airways in the lungs. However, the relationship between the two was not hypothesized on the basis of historical records, but discovered in a medical search of modern families carrying the CF mutant gene. Michael Swift and his colleagues at the New York Medical College reported in 1995 that by following up on anecdotal evidence they discovered that people who carry one copy of the most common CF mutant gene, the deletion at 508, were much less likely to suffer from asthma.

The numbers are clear: in extended families comprising people who have CF, relatives or siblings who carry one copy of the gene and relatives with no mutations, the incidence of asthma among those with one copy of the mutant gene was much less than the statistics would have predicted. So while it was expected on genetic grounds that nine or ten asthmatics out of twenty-five blood relatives would carry the 508 mutation, only four did so. Is asthma a reasonable candidate for the mystery disease that the CF gene is protecting against? Swift offers no specific cellular basis for such protection (and indeed there have been no significant respiratory changes noticed in people who have one mutant CF gene), but simply suggests ways that this hypothesis could be confirmed in larger studies. And while he points out that respiratory diseases are an important cause of infant mortality in underdeveloped countries, the distribution of the CF mutants must again be taken into consideration. For what reason could protection against asthma be limited to Europe?

As I was writing this chapter, Gerald Pier and his colleagues in Boston added typhoid fever to the list. They found that the bacterium causing typhoid fever uses the salt channel protein (in its normal form) to gain access to intestinal cells in mice. Furthermore, the number of bacteria invading those cells in mice carrying one mutant gene was reduced 86 per cent; mice with two mutant genes (the equivalent of having cystic fibrosis) were completely immune to attack by typhoid bacilli. While these experimental results might sound similar to those obtained for cholera, Pier's team argues that the recent appearance of cholera in Europe (see above) doesn't square with the ancient history of the CF mutant gene. However it is not clear that history is on Pier's side. There are few specific references to typhoid earlier than the seventeenth century; more problematic is the fact that typhoid usually spreads from a contaminated water supply and is therefore prevalent where populations are settled and concentrated. But if the mutant CF gene is at least 52,000 years old, it predates cities by tens of thousands of years. How were the hunting and gathering humans of the time being exposed to typhoid fever against which the mutant CF gene was protecting them?

That's where the mystery stands today. Everyone agrees that having one copy of the CF gene must confer some advantage, but against what still isn't clear. Cholera and typhoid fever are the best bets because they are backed up by some experimental evidence, but I wouldn't count out the possibility of a surprise candidate. The fact that there have not been experiments demonstrating a link between CF and candidate diseases like tuberculosis or influenza doesn't eliminate those possibilities. In tuberculosis, for instance, immune cells can build a fibrous container around the TB bacteria, preventing their spread through the lung. It has been suggested that the tendency of cells carrying the CF mutant to produce large amounts of viscous sugary materials might help to create such a protective capsule—but so far just suggestions.

The search for this mystery disease, and the concurrent exploration of the history of the cystic fibrosis mutant gene aren't simply exercises in academic curiosity. If there is a relationship between CF and cholera, for instance, it might be possible to use that knowledge to develop new cholera treatments. If closing chloride channels (rather than removing them) could offer protection, drugs that accomplish that might be new weapons in the fight against all childhood diarrheas.

Maybe as important in the long run is the fact that this case has opened researchers' eyes to the possibilities of gathering data on the prevalence of disease genes, mapping them and reconstructing their history. If it can be done for the deletion at 508, it can be done for others.

A Silver Lining

The more we learn about life on earth the more important seem the mechanics of survival. Brute force, the old idea of "Nature red in tooth and claw," is no guarantee of survival—living things must be subtler. There is no doubt that competition is everywhere, but contrary to popular opinion nature isn't so much a battle ground as a marketplace, where deals are made and swindlers stand next to honest brokers. Sometimes in the natural world, as in society, it's hard to tell the two apart.

Males display their tail feathers or antlers or croak deeply for females; other less-impressive males try to beat the system by sneaking a mating or two, while the ostensibly faithful females are often doing some sneaking of their own. (I guess I should make clear here that I am referring to the natural world *aside* from humanity.) One of the most intriguing strategies for getting ahead in life—at the expense of others—is something used by a variety of birds called brood parasitism: lay your eggs in another's nest and let them worry about providing for the offspring.

A Silver Lining

There are dozens of species that practise brood parasitism, the most familiar being cuckoos in Europe and cowbirds in the Americas. It sounds as though it should be a pretty straightforward game: the parasite seeks to deposit as many eggs as possible into the nests of unwitting hosts, while the hosts do their best to keep foreign eggs out and/or to avoid raising offspring other than their own. But the details of their relationship take some fascinating twists and turns.

The behaviour of the North American brown-headed cowbird is a good example of the to and fro that takes place. Female cowbirds have a kind of nine-to-five schedule—which immediately tells you they don't have offspring to raise. In the morning they lay one egg in some other bird's nest, often small songbirds (up to forty over the breeding season), then spend the rest of the morning looking for other target nests. Afternoons are devoted to feeding themselves. The males more or less follow the females around, although they take no part in the search for nests.

Raising a baby cowbird is usually a disaster for the host. The female cowbird begins by tossing out one of the host's own eggs. As time passes, the cowbird nestling matures faster, demands and gets the lion's share of food brought back to the nest, and may even crowd the remaining nestlings out of the nest. When you consider that the name of the evolutionary survival game is to get as many of your genes (as contained in your offspring) into the next generation, rearing cowbirds at the expense of your own is decidedly a bad thing.

Songbirds in North America have adopted a variety of strategies to combat the cowbird. Some, like yellow warblers, respond to the appearance of a cowbird egg in the nest by building a second nest on top and laying a whole new round of eggs. The problem with this strategy is that the cowbird might return to lay a second egg in the second nest, leading to a third and sometimes even a fourth nest. Obviously this is a losing strategy if the summer is spent making nests and laying eggs that never hatch. Other birds puncture the cowbird egg, or bury it, or even abandon the nest completely. One species, the blue-grey gnatcatcher, has been seen dismantling the original nest and carrying it, piece by piece, to a new location, minus the offending cowbird egg. Of course there is no guarantee that the cowbird will not find the reconstructed nest.

As one-sided as the relationship is, it sometimes takes on an even more malevolent flavour. Great spotted cuckoos in Spain lay their eggs in the nests of magpies, pretty tough and adaptable birds in their own right. But unfortunately for the magpie, the cuckoos' intervention doesn't stop there. They monitor the nest after they've laid their egg and will lay waste to it and the host bird's nestlings if the cuckoo egg is kicked out. In other words the host bird had better accept the cuckoo nestling—and the problems that creates—or it might well lose its entire brood. Animal behaviour specialists have called this the Mafia hypothesis, the idea being that cuckoos act as extortionists, "persuading" recalcitrant magpies to toe the line or else.

The great spotted cuckoo might destroy a magpie nest to persuade the magpie to engage in another round of egg-laying so the cuckoo can deposit a second egg in the nest, but researchers in Spain have some evidence that cuckoos will waste magpie nests even at the end of the breeding season when there's little likelihood of another brood. In cases like that it looks as if the cuckoos are teaching the magpies a lesson—a lesson that will probably be remembered next breeding season.

However, evolution is dynamic. No strategy, however airtight it might seem, will work forever for either participant. Hosts will figure out ways to avoid raising some other bird's nestlings, and cowbirds and other brood parasites will inevitably develop a response to these tactics. Sometimes it seems as if we humans are witnessing this dynamic while it's still in development. For instance, in England cuckoos of a different sort have chosen to specialize rather than dabble in organized crime. There are four versions of this cuckoo and each lays its eggs in the nest of a particular victim. Three of those unwilling recipients are vigilant in defence of their nests, rejecting any egg that looks different from their own. In return, the cuckoos lay eggs which resemble the eggs of their hosts, a response that has presumably been evolutionarily fine-tuned over millennia. They also take care to lay eggs after the host bird has laid some of her own, but not too long after, and the female cuckoo even chooses her hour of laying to ensure that the host birds are unlikely to be around the nest.

However, one of the four birds parasitized by cuckoos, the dunnock, seems incapable of recognizing eggs of a different colour than its own, or even whole clutches of them. Some biologists suspect that the dunnock is only a recent victim of the cuckoo, recent in this case being the two or three thousand years since the forests of England were cleared, opening up the countryside breeding habitat favoured by the dunnock. If that's true, it could be too early in the to and fro of host and parasite for the dunnock to have evolved the egg discrimination abilities sufficient to reject cuckoos' eggs. Interestingly the cuckoo that parasitizes the dunnock is the only one of the four in England that lays eggs that don't resemble the host egg, probably because it doesn't have to.

If that's an example of the arms race in its early stages, then in Africa there's a lovely example of the arms race taken to the nth degree. There, parasitic widow birds not only lay eggs that resemble the eggs of their finch hosts, but those eggs hatch to produce nestlings that have exactly the same patterning of coloured mouth markings as baby finches do. When the nestlings mouths gape open to stimulate the adult's feeding response, even a discriminating finch can't tell what species of maw he or she is gazing into. But clearly the implication is that the host parents would get rid of any nestling that didn't look precisely right, a degree of discrimination that the host birds in England have not yet attained—they'll feed any nestling at all.

It's a fascinating notion that we might have stumbled upon the cuckoos and their hosts in the middle of an evolutionary arms race. But intriguing as it is there are those who don't buy the idea. Amotz Zahavi, an iconoclastic observer of animal behaviour, has focused on the relationship between the cuckoo and one of its hosts, the reed warbler, to argue that there is no arms race happening. Some reed warblers toss out cuckoos' eggs, some don't. Zahavi claims there's a much simpler explanation for this inconsistency than to presume we are seeing an evolving but still incomplete retaliation by the warbler. He quotes research from Japan which shows that the explanation lies in the age of the host bird. Each reed warbler female lays eggs with unique markings. New reed warbler mothers aren't yet familiar with these patterns, and so would be foolish to gamble by tossing out eggs which look foreign. But experienced mothers do know their own eggs and will toss out those they know not to be their own. So perhaps it isn't an arms race, just a case, as Zahavi says, of "each individual does the best it can to succeed in reproduction."

All of this is background for a story that is one of my all-time favourites. It is told in a scientific paper called "The Advantage of Being Parasitized," published thirty years ago by biologist Neal Griffith Smith. He was working at the Smithsonian Tropical Research Institute in Panama, observing the behaviour of a brood parasite called *Scaphidura*, the giant cowbird.

Scaphidura lays its eggs in the colonial nests of two birds, oropendolas and caciques. At first glance the relationships among these birds seemed straightforward. Most *Scaphidura* females laid eggs that looked like the eggs of the host birds. But there was one outstanding anomaly. One type of cowbird curiously mimicked nothing at all.

Then Griffith Smith took note of the habits of egg-laying cowbird females. There were two diametrically opposed behaviours. Some females were covert: they skulked around the nests of their hosts, waiting until the host female was absent before laying their one egg (which mimicked the host eggs perfectly). Then there were the females Griffith Smith called "dumpers." They travelled around the nesting colonies in raucous groups, completely obvious to their potential victims, sometimes even driving the host female from the nest to lay eggs. Not just one egg: two, three or even five at a time.

The dramatic differences among the cowbirds were mirrored by their hosts, the oropendolas and caciques. They fell into two groups as well: either discriminators or nondiscriminators. As the names imply, they either tossed out the cowbird egg or they didn't. So here was a mystery: two kinds of cowbirds (or at least, one kind behaving in two different ways) frequenting colonial nesting sites in which there were two kinds of response by the host birds. How come?

The secret lay in an unexpected direction: the location of the nests. Oropendolas and caciques often built their nests close to or even encircling nests of wasps and biting (not stinging) bees. Griffith Smith noted that even the noise of these thousands of insects was threatening enough, in his mind, to deter predators like opossums. But these insects had another, subtler benefit to the birds. For reasons Smith was unable to determine, the presence of wasps and bees sharply reduced the numbers of a parasitic botfly called *Philornis*, an unpleasant creature which lays its eggs on nestling birds. The larvae hatch and burrow into the chick's body to feed. On average, any chick infected by more than seven larvae died.

In nesting colonies without bees or wasps, flies were everywhere and mortality high. In colonies built among wasps and bees, Smith found only rare larvae and no adult flies at all. Although

Smith was unable to determine what was going on between these insects, he did recover ninety-four fly wings from directly underneath wasps' nests over a period of two years.

Botflies, wasps, oropendolas, caciques and cowbirds are the major players: here is the connection among them. Whenever the colonial host birds' nests were built in the absence of wasps or bees, nestling birds often fell prey to botfly infestation, with one important exception. If a nest was home to a cowbird baby as well, the botflies had little or no impact on the nestlings. The numbers are fantastic: in nests where there was no cowbird, 382 out of 424 nestlings were parasitized. In nests with a cowbird, only 57 out of 676. The reason was simple: nestling cowbirds preened their nestmates, eating the larvae or pupae of the flies. Host nestlings ate nothing except what their mother brought to the nest. Even some of the 57 cases of parasitism in the presence of a cowbird can be explained: sometimes the cowbird egg was laid after the others, and hatched too late to save them from flies. Or sometimes the cowbird simply couldn't reach all the other babies in the nest.

In these cases then, it was better for oropendolas and caciques to have cowbird babies in their nests when there were no wasps or bees about. They might lose one nestling to overcrowding, but at least they wouldn't lose them all. This in turn allowed the cowbird females to parade around the nest colony, confident that their eggs wouldn't be rejected.

Of course, if the host birds' nests were protected by bees and wasps, then the additional labour of raising a cowbird baby wasn't worth it, and host females would get rid of an egg they recognized as foreign. In this situation, female cowbirds kept to themselves, waited for an opportunity, and laid an egg that looked just like the ones already in the nest.

You could ask, why bother with cowbirds at all? Why didn't oropendolas and caciques always build their nests around bees and wasps? As always, no strategy is airtight. If the bees and wasps built their nests late in the egg-laying season, those birds who had waited for them ran the risk of losing their nestlings in early rains. Sometimes the wasps or bees deserted their nests, leaving the birds without protection. And why did cowbirds engage in the risky business of laying eggs in the nests of discriminators? They too derived some protection from the ants and bees, both against the flies and other potential predators.

The irony here is, despite the complexity of the relationships and the uncertainties surrounding either strategy, colonies of oropendolas and caciques near bees and wasps (and free of cowbirds) were actually no more successful at raising young than those encumbered with baby cowbirds. Sometimes a parasite is a good thing.

How Things Work

Tee Time at the
Royal Institution

―――――――
―――――――

It may be true that the public has greater access to science than
ever before with reports in newspapers, magazines, radio, television and books. However, I don't think any of these has the same
appeal and impact as the most popular nineteenth-century venue,
the public lecture. Of course the public science lecture is not extinct,
but I doubt that the modern version compares with those in nine-
teenth- and early twentieth-century England in which some of the
greatest living scientists talked about their work, accompanied, not
by overheads or slides, but by actual experimental apparatus.

Tickets for the popular lectures by the great chemist Michael
Faraday were as sought after in the 1840s as tickets for the opera or
the art shows at the Royal Academy in London. The Faraday tradition was maintained by a series of scientists for at least another
century. In the late 1920s Sir Arthur Eddington began a series of
such lectures of his own by contrasting the two tables on which he
wrote his lectures. One was "substantial," a table of substance, the

kind of table that exists in our everyday lives. The other shadowy version was the "scientific" table, a table that supported the paper and pushed back against the tip of the pen, not because it was "solid" but because it was composed of unimaginable numbers of electrical charges swarming around and battering the paper and pen. He pointed out that it was these charges that prevented his putting his elbow through the table, although, he noted, it was more accurate to point out that it was not impossible to do so, just highly unlikely.

Maybe the passage of time makes these lectures seem more fascinating than they actually were, but I doubt it. The late 1800s and early 1900s were revolutionary times in science, especially physics, and public lectures provided an insight into that excitement combined with a performance by one of the personalities involved—what could be better?

I would love to have been at Eddington's lecture, but it would be my second choice. If I could attend only one, it would be J.J. Thomson's lecture to the Royal Institution, in London, on Friday evening, March 18, 1910. I would choose this lecture for two reasons: it offered a unique blend of science and entertainment (a mix we could use more of today) and it was a lovely example of a great scientist's ability to demonstrate the unity of nature, the parallel between the great and the small. In that lecture Thomson used a cathode ray tube to explain the dynamics of . . . a golf ball.

J.J. Thomson is known today as the man who discovered the electron. Electrons are the negatively charged particles that move about in the outer spaces of the atom. Unfortunately at some point in our education most of us acquired a completely inaccurate mental picture of electrons as analogues of planets in the solar system. In this scenario, as planets orbit the sun, electrons orbit the nucleus of the atom. Not true. Electrons do spend most of their time outside the nucleus (although it's not impossible for them to penetrate it), but they do not occupy well-defined orbits. Instead they are found in vague probabilistic locations, best illustrated by clouds—clouds of probability. It's impossible to predict exactly where a given electron is or where it's headed, and if you try to determine one or the other, you preclude the possibility of knowing both.

This quantum mechanical picture of the electron was unknown in the 1890s. In fact the atom was still thought by most to be indivisible, which, after all, was why it was called the atom in the first place, from the Greek *atomos*, meaning "uncuttable." It was the ultimate particle. But by the beginning of the 1890s, some peculiar observations had begun to raise questions about the nature of the atom; many of these had been made possible by the use of cathode ray tubes.

These were the forerunners of the cathode ray tubes found in television sets. Turning on the power of your television set heats a metal filament so that electrons begin to boil off its surface—that is the cathode. This stream of electrons is focused onto the television screen and (very rapidly!) scans the screen, moving back and forth, up and down. The screen is coated with materials that glow under the impact of high-speed electrons, and pictures are the result.

The cathode ray tubes of the 1890s were essentially the same design; the big difference was that nobody really knew how to explain what was going on inside them. A typical cathode ray tube was made of glass with as much air as possible evacuated from it. When a metal filament at one end was heated to red-hot, a luminous beam materialized, extending from the tip of the filament like a flashlight beam in the fog. Where the beam met the glass at the end of the tube a patch of light appeared. The better the vacuum in the tube, the less visible the beam, but it was easy to show that it was still there, even if invisible. A metal object in the path of the beam would heat up enough to give off X-rays.

The Barmaid's Brain

There was a controversy over the nature of the beam. It was clear that it carried negative electrical charge, but what exactly was it made of? English scientists believed that it was a stream of particles, at least partly because it could be deflected by bringing a magnet—even a simple horseshoe magnet—close to it. That can't be done with a beam of light. Yet German scientists maintained that the cathode ray beam was more like light. In their favour was the fact that the beam apparently couldn't be bent by passing it between the positive and negative poles of a battery, something that should have been possible if it were a stream of electrically charged particles. The Germans explained the magnetic deflection of the stream by arguing that the magnet deformed, not the stream itself, but the medium through which it as moving. A modern analogy might be the Einsteinian view that the planets don't circle the sun because they're being tugged by the invisible force of gravity, but instead because space is warped by the sun's mass and the planets are simply rolling along, as if in furrows.

J.J. Thomson strongly suspected the cathode rays to be streams of negatively charged particles, but he needed to be able to demonstrate that conviction experimentally. The door was opened when he was able to show that the Germans' inability to deflect the beam by applying electric fields was simply the result of the tube being inadequately evacuated. Once he had accomplished that, he was away. He was able to exploit the fact that the charge, the mass and the velocity of these hypothetical particles would all be interrelated—for instance, the more massive the particles, the less easy it would be to bend the beam in any given direction. He could calculate the values of each by varying the strengths of both magnetic and electric fields and recording their effects on the beam.

Thomson had earlier discovered that the velocities of the mystery particles were much lower than the speed of light (another blow to the German contention that the stream was light-like). He now discovered that, on the other hand, the velocities were much higher than the velocities already measured for molecules and so presumably the particles were much lighter. In support of that notion he found that these particles had much less mass for every unit of electrical charge—*much* less. In 1897 Thomson announced —in a lecture, what else—that these things, which he called "corpuscles," were a thousand times lighter than the hydrogen atom, which at that time was the smallest atom known and so considered to be the smallest unit of matter.

Apparently when Thomson made this claim at the Royal Institution on April 30, 1897, the audience didn't exactly leap from their seats. Then he went on to argue that what was happening in the cathode ray tube was that atoms were being torn apart, releasing these negatively charged bodies (then "corpuscles," now electrons). When scientists heard these claims of Thomson's they greeted them with incredulity. One in particular argued that if it were true that all different kinds of matter contained identical "corpuscles," then the ancient alchemist's dream of transmuting one substance into another should be possible. Regardless of what his peers thought at first, Thomson was right and he received the Nobel Prize in 1906.

So if you were attending the Royal Institution's regular Friday evening lecture on that March day in 1910, you were looking forward to hearing one of the greats of twentieth-century science in his prime, a man who might well be expected to address fundamental questions of science and philosophy, like the nature of matter. But golf? What possible connection could there be between a golf ball and a cathode ray tube? Leave it to J.J. Thomson to find a way.

"There are so many dynamical problems connected with golf that a discussion of the whole of them would occupy far more time than is at my disposal this evening." So he began. In the first half of the lecture Thomson stuck to relatively straightforward aerodynamics, explaining how the spin of the golf ball affects its flight. With the ease of someone who is comfortable with his subject, he suggested the audience focus on the "nose of the ball."

A golf ball, said J.J., or any other ball in the air, will follow its nose, the most forward point on the ball's surface. This holds true no matter how the ball is rotating: if a ball has backspin, that is, it is rolling backward in the air, the nose is constantly rising, and so the ball will rise also. If a ball in the air is turning clockwise (as seen from above) then the nose will forever be rotating to the right and the ball will do so as well. That is a good description of a badly sliced drive. Golfers are all too familiar with "topping" the ball, giving it overspin. In that case the nose of the ball is turning down, and the ball quickly follows, usually to the amusement of several onlookers standing idly by the first tee.

Having established that spinning golf balls change direction, Thomson turned to experiments to demonstrate why. One of the most ingenious involved two golf balls, one with dimples and one without, that had been skewered on a metal rod like chunks of lamb on a shishkabob. The metal rod was mounted vertically on a stand and connected to an electric motor, so that the rod and the two golf balls could be spun rapidly. A powerful fan stood nearby—when switched on it would send a blast of air over the balls.

Thomson was about to demonstrate two things: one, that a spinning ball curves because of variations in the movement of air over its surface, and two, that dimples make a difference. To do this he needed one other piece of equipment, a U-shaped glass tube with liquid inside, the open ends of which could be brought close to the opposite sides of a spinning ball. If there were differences in pressure from one side to the other, the liquid would move inside the tube. Thomson first showed that as long as the balls weren't spinning, no matter how strong the blast of air over them, there were no pressure differences from one side to the other.

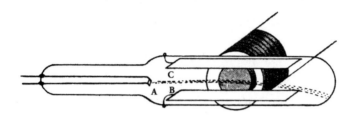

*This cathode ray tube may look like a slightly antiquated
piece of scientific equipment—and it is. But in the right
hands it becomes a miniature golf course.*

However, as soon as the balls were being spun by the electric
motor, the liquid shifted position in the tubes, proving that the air
from the fan had created noticeable pressure differences from one
side of the ball to the other. It was also clear that the dimpled ball
created more than twice the pressure difference the smooth one
did. Thomson's point was this: the reason that a spinning ball
"follows its nose" is that on one side, the air rushing past the ball is
travelling in the same direction as the ball's surface, while on the
other side the ball is actually turning *into* the oncoming air. On
the side of the ball that is turning against the flow of the
oncoming air, pressure is raised. On the other side, where the spin
and the airflow are in the same direction, pressure is lowered. The
ball, sensibly enough, heads off in the direction of lower pressure.
Dimples on the surface of the golf ball enhance this effect, which
is a good or bad thing depending on what kind of spin you put on
the ball.

To this point in J.J. Thomson's lecture you might have been
impressed with the ingenuity of his experimental apparatus and his
lucidity, but the best was yet to come. He now produced a cathode
ray tube (the last piece of equipment you'd expect to see in a
public lecture about golf) and proceeded to draw an analogy

between the flight path of a golf ball (not the ball itself) and a beam of electrons. At one end of the glass tube there was a red-hot platinum filament emitting electrons which were plainly visible as a luminous streak running the length of the tube. The platinum was the tee; the streak of electrons, the path taken by the ball.

Thomson then announced he was going to apply the "force of gravity" to the flight of the ball by connecting a battery that supplied electricity to two plates inside the tube. The plate above the flight of the ball was negatively charged; because the electrons are also negative charges, the plate repelled the electron stream, making it bend downwards as it travelled the length of the tube, just as a golf ball falls back to earth.

But that was just the beginning. Thomson then applied a magnetic force to the electrons, arguing that the effects of spin on a ball and magnetism on an electron stream are mathematically equivalent. Even if you know nothing of the mathematics, you can discern the similarities in the language physicists use:

The spin of a ball produces a force that acts at right angles to the direction of motion, right angles to the axis of spin, and is proportional to the product of the velocity of the ball, the velocity of spin and the sine of the angle between the velocity and the axis of spin.

A magnetic field produces a force on electrons that acts at right angles to their direction of motion, right angles to the magnetic force, and is proportional to the product of the velocity of the particles, the magnetic force and the sine of the angle between them.

So now J.J. went to work, turning on the electromagnet and deflecting the stream of electrons (the path of the golf ball). First, backspin. The ball carries farther than it did when just under the influence of gravity. Then more backspin, more lift, higher altitude, greater carry. Finally there's so much spin on the ball that it rises sharply then slows and falls nearly straight down. But still Thomson doesn't stop, and with even more electromagnet, more spin, the golf-ball-in-a-tube does a loop-the-loop. More yet, and the ball loops its way down the tube, three, even four loops from one end to the other, a golf shot that Thomson notes, "we have to leave to future generations of golfers to realize." He does point out that he could apply even more spin, making it possible for the ball, once struck by the club, to rise up over the tee and land behind the nonplussed golfer.

I have never understood just why J.J. Thomson chose to use his beloved cathode ray tube to explain the flight of a golf ball. It might have been his love of the piece of apparatus that enabled his great insight into the nature of matter; it might have been the attraction of the mathematical equivalence of electron beams and the flight of golf balls. Apparently it wasn't his skill at the game. He did play, but there's no evidence he played particularly well. But in giving this lecture he was taking his place alongside another great English physicist—in fact, the greatest of them all, Isaac Newton.

There is no way Isaac would ever have bothered to give a public lecture—he never gave any sign that he cared what the public thought of his work—but in the early 1670s he wrote to an acquaintance that he had watched tennis players make the ball curve with a sideswiping motion of the racquet. Newton made it clear he understood that the cause was differences in pressure on the two sides of the ball, exactly the principle demonstrated by J.J. Thomson at the Royal Institution on that Friday night. And I would like to have been there.

It'll Practically Go Forever

C.P. Snow, the prolific novelist and scientist who coined the term "two cultures" to describe the split between the arts and the sciences, recalled in his book of the same name the frustrations he experienced attending gatherings of educated people:

> Once or twice I have been provoked and have asked the company how many of them could describe the Second Law of Thermodynamics. The response was cold: it was also negative. Yet I was asking something which is about the scientific equivalent of: *Have you read a work of Shakespeare's?* (italics his).

He wrote those words in 1959; today the question should probably be rephrased: "Have you even *heard* of the Laws of Thermodynamics?" or "Can you *spell* Thermodynamics?" There are many people (some I've met who apparently take pride in knowing nothing about science) who would be perfectly able to argue the value of having read Shakespeare but would see no usefulness at all in being aware of chemical laws. I like to point out that while it's true that such laws might not make it possible to increase your RSP earnings, they both describe the universe we live in and reveal the mysteries still contained in it. But sometimes this isn't persuasive enough and I need to pull a more practical reason for knowing something about science out of the hat. Happily I have one in this case: if you are familiar with both the First and Second Law of Thermodynamics, you will be much less likely to waste money investing in a perpetual motion machine.

The perpetual motion machine is one of the most entertaining fantasies of the unscientific kind (and one that can boast one of the longest lineages). By this term I don't mean machines that use renewable resources, like solar, wind or tidal power—I mean machines that are supposed to create energy out of thin air. The typical perpetual motion machine may need a little something to get it started: the turn of a hand crank, the juice from a small battery, but once it's going, it goes forever, using the energy it generates both to do work *and power itself.* It is the age-old promise of something for nothing, clothed in technology.

There are several centuries' worth of perpetual motion machines and the intriguing thing about all of them is that as the science on which they were based evolved—and showed why they were impossible—new designs were created that once again pushed beyond the existing knowledge. And so it has gone, century after century. But while some of the more recent machines were plainly designed to fleece a gullible public, the earlier versions were charmingly straightforward inventions based on the belief that nature will really surrender something for nothing. But first, before I introduce them, a little of the relevant science. But I warn you, even armed with the science, you will find the lure of perpetual motion hard to resist.

There are three laws of thermodynamics, but the first and second are the important ones in this case. The first says that energy can neither be created nor destroyed, an important concept when faced with a perpetual motion machine, which must, by definition, generate more energy than it uses. This energy has to come from somewhere. Another way of stating the First Law of Thermodynamics is: "You can't win." This does not mean that energy cannot be changed from one form to another. An important sidebar to the first law is that when work is done, a certain amount of energy is converted to heat. It happens under the hood of your car, when you run upstairs, even when you sign your name on a cheque. In the first two cases it's obvious, but even the heat produced by the friction of the tip of your pen on the cheque, although minuscule, is measurable.

The Second Law of Thermodynamics, roughly stated, is that energy cannot run uphill. The heat that is produced in the pen as you sign your name can't be turned around and used to move your fingers. It can, however, be used to help melt a tiny ice-cube, a good illustration of the restriction that heat can only do work if it flows downhill, like water. Heat will always flow from a higher temperature to a lower one and it can do work as it goes. However, once it has reached that lower temperature, it is stuck there, its only option being to cool off even more. The pen will melt the ice-cube and simultaneously lose some heat; the ice-cube will not grow and by so doing heat up the pen.

This leads to the startling conclusion that while energy is neither being created or destroyed (the first law) it is being converted gradually to heat, which itself is busy being redistributed from warm places to cold, the ultimate end being that all energy will have been used to make everything in the universe the same temperature, and no work or machines of any kind will be possible. This has been called the "heat death" of the universe, and while it is obviously something we don't have to worry about in the near future, it is a powerful example of the Second Law of Thermodynamics at work. If the First Law of Thermodynamics can be rephrased as "You can't win," then the second is, "You can't even break even."

Heat means something specific to a physicist: it is a measure of the movements of atoms and molecules. The hotter they are, the faster and more violently they move. It is a statistical measure: for instance, it is unlikely that all the molecules of air in a balloon are moving at the same speed. In that sense the heat of the balloon is a composite figure derived from all the molecules inside. It is conceivable, although highly unlikely, that for a moment all of the fastest molecules could find themselves, by chance, in the same area of the balloon. That small area would then be much hotter than the rest and for a moment the second law would have been defied. But only for a micromoment.

This example shows that these are not inviolate laws; they are probabilistic descriptions of the movements of atoms and molecules. However, when proponents of perpetual motion machines try to argue that it is exactly these loopholes that their machine is exploiting, it's worth a closer look at that word "probabilistic." A chemist named Henry Bent once calculated that the likelihood that one calorie of heat energy (the amount needed to raise one gram—roughly one millilitre—of water one degree Celsius) could run uphill and be converted completely into work (reversing the Second Law of Thermodynamics) was the same as if that famous gang of randomly typing monkeys produced the complete works of Shakespeare 15,000,000,000,000,000 times in a row—without error. Run your perpetual motion machine on *that*.

On to the machines themselves. There have been thousands designed, many built, but exactly zero working examples produced. My favourites are those from the Middle Ages and the

Renaissance, mostly because at first glance they seem perfectly reasonable. One of the first perpetual motion machines on record was designed by the Italian Mark Antony Zimara in the early 1500s. There is no evidence it was ever built; Zimara apparently never even had drawings made of it. He imagined a windmill built close to a set of huge bellows, powerful enough to move the vanes of the windmill with their blast. The bellows of course had to be squeezed together to provide that wind and Zimara therefore added a third crucial component to his invention: an "instrument" that connected the two and allowed the turning of the windmill to compress the bellows. He wasn't very specific about exactly how that would be accomplished, but you could imagine a set of levers that would translate the rotary motion into a horizontal squeezing force.

It is a clever design at first glance. However if you tried to build this machine you would quickly discover that the force needed to work the bellows is much greater than anything that would be produced by the windmill, with the result that the machine would stand completely motionless, even the potentially useful rotation of the windmill stalled by the bellows.

Water, not wind power, was the favourite motive force for designers of perpetual motion machines. A classic design (that was repeated over and over through the centuries) was based on the water-mill. By the fifteenth century almost every English manor on a stream had its own mill. A normal water-mill featured a large wheel, not unlike those that powered the Mississippi riverboats (although much slimmer), turned by the flow of water in the stream. The rotation of the wheel was converted by gears into the powerful grinding movements of a huge stone mill. In 1618 the English physician Robert Fludd gave this basic design a neat new twist. He added something called an Archimedean screw, named of course after one of the greatest mathematicians of all time, Archimedes (see "The Burning Mirrors of Syracuse"). The Archimedean screw

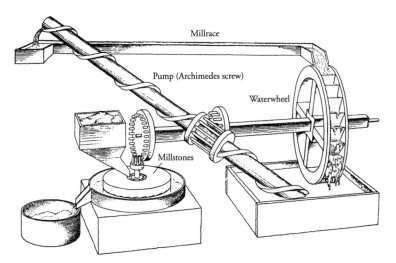

This seventeenth-century perpetual motion machine makes perfect sense doesn't it? The water goes 'round and 'round, driving the millwheel and itself. But of course it doesn't work.

is a pipe that is threaded on the outside. If such a pipe is immersed in water and turned, water will be caught on the threads and gradually carried to the top of the pipe. To Robert Fludd, the next step must have seemed obvious. Disconnect the wheel from any outside source of flowing water and arrange the screw and the mill so that water flows around and through both. His design has water flowing along an elevated millrace, then down over the wheel, turning it as it falls, finally splashing down into a trough where it is picked up by the Archimedean screw and lifted back up to the top of the wheel again. The water moves the wheel, which turns both the Archimedean screw and the millstone. Surely the force involved in turning a millstone could run some water uphill. It's brilliant! It's perpetual! It doesn't work!

Fludd had no way of knowing that chemical laws would be formulated two centuries later that would show why his scheme couldn't work (although that doesn't excuse those who were still bringing this identical concept to the US Patent Office in the late-nineteenth century). His scheme wouldn't fly because the work necessary to raise the water back up to its original level is equivalent to the work generated by the water turning the wheel—but a good part of that work has been used to turn the millwheel. There's not enough left to turn the Archimedean screw and furthermore, much of the energy is lost through the friction of the water against the walls of its container and the wheels turning on their axles. So even if there were no millstone, extra energy would have to be put in to raise the water up to the top again. But like most perpetual schemes, it looks pretty good on paper.

Overbalanced wheels have been another favourite design. In these, a wheel is decorated by a series of movable weights attached to its perimeter. Sometimes they are at the end of short mallets, sometimes the weights roll in and out along the spokes of the wheel—the exact design doesn't matter. The idea behind them all is that at any moment as the wheel turns, the weights shift position in a way that ensures there are always more weights on one side than the other, so the wheel is compelled to continue to turn.

These are again convincing to the eye, until you notice that in every case a weighted arm has flipped over from one side of the wheel to the other before it actually would, or a ball is not rolling along a spoke when it actually would be. In reality, in all such wheels that are actually built, the locations of the weights at all times are symmetrical and the wheel will not turn by itself. You could, of course, keep flipping one of the top weights over with your finger to keep the wheel turning. By adding this small amount of energy you would be balancing the books of the first law.

Another simple beauty was dreamt up by a Jesuit priest named Johannes Taisnerius in 1570. This was based on the then mysterious force of magnetism, which at first glance does appear to

provide a never-ending source of power in exchange for nothing at all. But the trick is to put that to work in a machine. Taisnerius's version combined a strong magnet fastened to the top of a post, with a sloping track running from the magnet down to the tabletop. A metal ball was placed at the bottom of the track, and, under the pull of the magnet, began to roll up the track towards the top. However, just before it reached the magnet, it dropped through a hole in the track and returned to the starting point via a second track. What could possibly be wrong with this?

Just this: if the magnet were strong enough to pull the metal ball all the way up the track, it would certainly be strong enough to pull it right over the hole. Even some experts in the 1600s who held out hope for perpetual motion realized that this magnetic system wasn't going to work. The fact that no practical application had apparently been planned for this device has led one modern commentator to label this as the first executive desk toy.

All of these machines were supposed to work with real-world materials. There are other lovely perpetual motion devices that would certainly work if their somewhat exotic ingredients existed. One such machine would rescue the now-disgraced overbalancing wheel from obscurity, providing the necessary anti-gravity material could be discovered. It consists of a simple wheel mounted on an axle with some sort of anti-gravity stuff placed under one half of the wheel. That half of the wheel is "weightless"; the other half feels the pull of gravity and the wheel turns in that direction. I imagine that when we discover how to neutralize gravity, turning the wheel on a perpetual motion machine will be far down the list of potential uses.

The examples I've cited so far were, as far as anyone can tell, genuinely believed to be feasible by their inventors. Harsh reality intervened only when the attempt was made to transform the pencil-and-paper version into an actual machine. But whenever it is possible to promise something for nothing, to make the impossible look convincing, you can be sure that shysters won't be far behind. One of the cleverest of those was a man named Paine who exhibited a perpetual "electro-magnetic" machine in Newark, New Jersey, in 1870. This was a powerful device that ran lathes and sawed wood, the only evident power source being four tiny (and completely inadequate) batteries. A skeptic named Dr. Henry Morton viewed the machine one afternoon and although convinced that it was fraudulent, couldn't figure out where the extra power was coming from. Then an odd thing happened: Paine tried to restart his machine to provide another demonstration, but it wouldn't work. He offered some reasonable excuse, but Morton noticed that it was 6:05 p.m. At 6:00 sharp, the end of the workday, the steam engine in the building that powered machinery in other rooms had been shut down. Morton was certain the two events were connected. Sure enough, when shortly afterward Paine left town suddenly and quietly, a search of the room where the machine had been demonstrated revealed a hole in the floor for a belt drive connecting the perpetual motion machine to the steam engine on the floor below. Remember the first law: energy cannot be created. Four small batteries need outside help in running a large machine.

In other similar cases the extra energy has been supplied by a hidden person turning a crank or by a current of air supplied through an unobtrusive pipe turning a wheel, but sometimes the illusion is more subtle. One of the most celebrated recent cases involved Mr. Joseph Newman who, in the early 1980s, tried to get a U.S. patent for an "Energy Generation System Having Higher Energy Output than Input." When it comes to applications for patents on perpetual motion machines, the US Patent Office has, for more than a century, refused to consider anything less than a working model. Newman objected to that restriction and a long court battle ensued. A visit to Newman's website (http://home.

earthlink.net/~josephnewman/) tells you more than you'd ever want to know about Mr. Newman's protests against conspiratorial government bureaucracy, his radio interviews and the current edition of his book, *The Energy Machine of Joseph Newman.* Mr. Newman might have been a hit on the "Tonight Show" (he says so) but scientists aren't quite as impressed. A back-of-the-envelope calculation by one physics teacher and his students after seeing a television demonstration of a car propelled by Newman's energy machine revealed that the car could have been moved at the very low speeds demonstrated just by the power of the hundreds of batteries that were apparently on board. Mr. Newman claims the batteries merely excite copper atoms, which in turn generate a magnetic field, providing the energy to power the motor. However, he has never submitted his "energy machine" for testing.

The only thing perpetual about perpetual motion is the unflagging efforts of believers either to persist in claims that their working models are being ignored (see above) or to search for novel ways of detouring around the laws of thermodynamics. This latter effort has time on its side: just as science advanced to show why the early machines couldn't work, it continues to move ahead, sometimes opening up new possibilities.

The latest efforts to get something for nothing have focused on quantum mechanics, the world of the atom and its constituent particles. The quantum world is a strange place—unlike anything we are familiar with—where common sense might as well be abandoned. Among its many counter-intuitive ideas is a rule that seems to have opened a tiny loophole in the laws forbidding perpetual motion machines.

That rule is Heisenberg's uncertainty principle, which says there is only so much you can know about the subatomic universe. For instance, if you determine the precise location of a particle, you cannot know its velocity, or vice-versa—measuring one feature precludes measuring the other. Knowledge is limited and everything is in flux.

This has startling implications for the vacuum of space, which, by its name, should contain absolutely nothing. But the uncertainty principle proclaims that even the vacuum cannot be completely empty. It is a sea of energy. (Even weirder is the idea that this energy can give rise to momentary "virtual" particles, pairs of minuscule objects that emerge spontaneously from the sea of energy, exist for almost no time at all, then mutually annihilate. These pairs typically last about one million-billion-billionth of a second.) This energy of the vacuum is the closest thing to getting something for nothing that physicists can imagine, and one or two of them, have, with true perpetual motion verve, taken this idea and run with it. They haven't gotten too far, but they *are* running.

One of them is physicist Harold Puthoff at the Institute for Advanced Studies in Austin, Texas. Dr. Puthoff claims in an article titled *A New Rosetta Stone in Physics* that "zero-point energy," as it's called, will revolutionize the physics of the twenty-first century (the name refers to the fact that the uncertainty principle establishes that this energy will be present even at absolute zero, where theoretically all molecular motion ceases). Apparently he has so far examined at least ten devices submitted to his institute which claimed to be able to extract energy from the quantum fluctuations of empty space, but none managed to extract more energy than was put in. Undeterred, Dr. Puthoff wonders if there will be "vacuum engineers" in the future and asks whether the energy crisis might be solved by "harnessing the energies of the zero-point sea." There is no doubt that zero-point energy exists. In fact, fluorescent light bulbs depend on it: in them, mercury atoms are first excited, then unpredictable blips of energy from the vacuum trigger the release of that excited state as light. But in this case the vacuum is just helping, not serving as the source of the light.

It'll Practically Go Forever

In seeing the energy of the vacuum as the something-for-nothing of the next century, Dr. Puthoff is ignoring the reservations of noted physicists like Nobel Prize winner Steven Weinberg (who argues that the amount of energy in a vacuum the size of the earth would be about as much as is contained in a gallon of gasoline), and skeptics who note that Puthoff and colleague Russell Targ spent much of their time in the 1970s arguing that ESP, psychokinesis and precognition exist and could be used to advantage at the Las Vegas gaming tables. The two were also convinced that spoon-bender Uri Geller possessed true psychic powers, an argument that would persuade very few these days. Puthoff, however, maintains that just as the twentieth century will be known as the nuclear age, the twenty-first will be the zero-point energy age.

My fearless prediction is that even if the energy of the vacuum can be tapped, it will prove to amount to something less than Harold Puthoff's dreams, and you know what that means. The twenty-first century will generate new and even more fascinating perpetual motion machines. All will be wonderful to look at and all will fascinate the public. Some who forget the laws of thermodynamics will even invest in them. None, however, will work.

Going Up?

Science fiction is full of scenarios for the future ("Beam me up Scotty) that never have to meet scientific scrutiny. It wasn't until Laurence Krauss wrote *The Physics of Star Trek* that anyone bothered to wonder whether warp drive, phaser beams or the holodeck were even theoretically possible, let alone practical. Still they were good ideas from the point of view of entertainment, but there are better: ideas that are as imaginative and intriguing as anything in Star Trek but also have some hope—however slim—of actually coming to pass. My favourite in that group is the "space ladder."

I suppose the notion of a space elevator has its intellectual roots in the Biblical Jacob's Ladder or even Jack and the Beanstalk, but both those extended from the earth to aerial realms that we know not to exist. On the other hand the modern concept of the space ladder has a starting point: communications satellites. Science fiction author and visionary Arthur C. Clarke was the first to point out, in 1945, that a satellite over the equator at the right distance from the earth—about thirty-six thousand kilometres—would orbit at just the right velocity to keep it above the same point on the earth all the time. It would appear to be hovering, although that is slightly misleading; a hummingbird uses a tremendous amount of energy to hover over a flower, but a "geostationary" satellite is racing through space at the same speed as the point on the earth directly below it is racing along carried by the earth's rotation. The satellite is in free-fall, but far enough from the earth that it falls around the planet, not into it.

Clarke pointed out that such a satellite would be able to beam signals halfway round the earth, and his foresight has been more than amply borne out (the problem today is finding spots for all the communications satellites countries wish to launch). But Clarke did not point out another potential of geostationary satellites: they could serve as an anchor point for a space ladder. After all they are permanently in place above the earth, so why not just let down a cable, Rapunzel-like, to the ground? Once that was anchored you could simply climb it—given a fair amount of time—into space.

It is a wonderful, why-didn't-I-think-of-that idea (first conceived by a scientist named Yuri Artsutanov and published in *Komsomolskaya Pravda* in 1960), but it has been taken far beyond that. Engineers and space scientists have wrestled with the practical difficulties of building a space ladder and have concluded that it would be an enormous task—roughly equivalent to building a suspension bridge around the world—and that there are a number of problems that would have to be solved, but that it might be possible.

First the physics. The fact that geostationary orbit is 36,000 kilometres above the earth makes this a challenging (some would say impossible) project. The problem is to prevent the space ladder from buckling or collapsing on itself. Early studies suggested that ten kilometres was the height limit for an aluminum tower, and even if you use incredibly sophisticated materials, much lighter and stronger than aluminum, you're talking here about building a ladder thousands of kilometres higher than that. It is true that the problem of collapse—the tower crushing itself under its own weight—is reduced because the gravitational pull on the tower lessens with altitude, meaning that the upper reaches are feeling the pull of the earth less. Taking that into account, the total weight of a space ladder turns out to be something less than 14 per cent of the weight if gravity acted uniformly throughout the height of the ladder.

But that is still the equivalent of building a 4,900-kilometre tower. If you were to start to build such a tower from the bottom up (the CN tower writ much larger), it will almost inevitably collapse on itself. The bottom of the tower would have to be mountain-sized, or it would be unable to withstand the enormous weight of the rest of the tower above. The engineering term is compression; the result would be collapse.

However, as aerospace engineer Jerome Pearson pointed out in a 1975 article, you don't have to build a space ladder from the ground up—it makes much more sense to build it from earth orbit down. Pearson envisaged a factory in orbit that would actually extrude two ladder-like strands of material at the same time, one heading down to the earth, the other heading straight up away from the earth. This is done to equalize the forces on the orbiting factory. As the ladder extends from geostationary orbit down towards the planet, it will begin to feel the tug of gravity more and more. At the same time, the strand extending into space will experience a corresponding increase in centrifugal force, the same as that experienced by the stone in David's sling. Both strands would be thin at first, but would gradually thicken to support the length of cable already deployed. If the outward strand were to be extended far enough, it would experience an outward force equivalent to the gravitational pull on the ladder, maintaining the orbital stability of the factory and also opening up interesting possibilities once the ladder has touched the earth.

This arrangement solves one huge problem: the ladder extending down to the surface of the earth is suspended from above: in engineering terms it is under tension, not compression, so it will not collapse under its own weight. At the same time it cannot pull the satellite cable factory down to earth because the outward cable is balancing it. The whole system is just . . . there. If the ladder is cut at the earth's surface, nothing happens. It just hangs there. If an airplane slices through it, that section below the plane will fall to the earth. The section above will react much more deliberately to the loss of weight by rising slowly to a higher orbit.

One problem may be solved but another has been created. Because the centrifugal force pulling out and the gravitational force pulling in vary differently with distance, the outward cable would have to be extremely long to balance the ladder and maintain the proper tension on the system. It would reach 110,000 kilometres into outer space from the orbiting factory, making the entire system about 144,000 kilometres long. That would be a metal strand extending a third of the distance to the moon. That upsets several scientists, including Arthur C. Clarke (who although failing to consider this in his seminal paper in 1945 is now a space ladder enthusiast). He argues for replacing the enormous mass provided by thousands of kilometres of cable with something like a captured asteroid, perhaps gigatons in mass, that could be positioned much closer to the orbiting factory and still provide the necessary counterweight.

Those are some of the engineering principles involved, and with those in mind, the concept can be fleshed out a bit. Of course it wouldn't be a space "ladder" in the sense of rungs and rails. Climbing one of those would be a taxing workout indeed, especially in the absence of oxygen at higher altitudes. The ladder would likely be an elevator, with cars moving up and down in some sort of open framework. A station for arrivals and departures would eventually replace the factory that spun out the cables in the first place. I mentioned that both the ladder and the corresponding space extension would taper; the ladder being thickest at the top and thinnest where it met the earth. Tapering in this case need not be literal; it is a question of mass of material and could be accomplished by building a hollow elevator shaft whose walls would thin gradually, the closer it was to the earth.

Notwithstanding the ingenuity that has been poured into this already, there are still two very difficult engineering problems to overcome before the space elevator becomes a reality. The first is the material to build it. Even with the removal of the compression problem by deploying a counterweight and the alleviation of the buckling problem by keeping the amount of material to a minimum, this is still an unearthly structure, and there are no materials in use today that would fit the bill. Much of the original thinking about the space ladder took place in the 1960s and '70s, when the best materials available (and just barely) were graphite fibre composites of the kind seen in tennis racquets and fishing rods. By the 1980s enthusiasts were talking about fibres comprised of perfect crystals of graphite, a form of carbon which would be enormously strong and just capable of building the space elevator. Of course that capability was dependent on scaling up what had been seen in the lab with tiny whiskers of this material to enormous cables.

None of the scientists making these projections could have imagined the discovery of buckminsterfullerene, another natural form of carbon, that just might make the concept of space elevators a little more respectable. Buckminsterfullerene is a cage-like molecule built from sixty carbon atoms, bonded together in the pattern seen on a soccer ball or in a geodesic dome (hence the homage to Buckminster Fuller, who, curiously enough, had his own take on space elevators—see halo-bridges below). This molecule is called a buckyball for short, and has, since its discovery a little over ten years ago, fascinated chemists. It is another fundamental form of that very versatile element carbon, as are graphite and diamond. But buckminsterfullerenes don't just come in spheres; they can also form buckytubes, and these are incredible molecules.

Their length can be literally millions of times their diameter, and they are constructed in a way that makes them almost impervious to tension (remember how important tension is in a space elevator). How thin are these buckytubes (sometimes called nanotubes because they are about one nanometer across, a billionth of a metre)? If you could construct a buckytube that would reach from the earth to the moon, you could roll it into a ball the size of a poppy seed. According to the experts you would only need to roll it "loosely" to make it that small.

This is incredible stuff, and it has some qualities that would make space elevator engineers take notice. If a bundle of buckytubes could be built (and right now that's still on the drawing board), that bundle would have enormous strength. A rope of buckytubes two and a half centimetres thick (a buckycable?) would be a hundred times stronger than an equivalent rope of steel, but only one-sixth the weight. It would also be resistant to breakage: a crack or fracture in any one of the countless tubes in the bundle would not affect the others, so damage would not propagate through the bundle as it does in other materials. That kind of strength puts the buckycable in the ballpark for building a space elevator.

Being able to construct the shaft for the elevator brings us a step closer, but there's still the matter of getting the elevator up the shaft and into space. A slow climb would be unacceptable to future humans who will likely be accustomed to getting places fast. But getting to the station in less than six hours would require the elevators to travel more than 6,000 kilometres an hour, ten times faster than a commercial jet. Speeds like that preclude the use of wheels or anything at all that would generate friction between the elevator and the shaft, so some sort of magnetic levitation system would have to be used. Most of the scientists who have considered this suggest that electric power would be the most reasonable way to move the elevators (although Arthur C. Clarke suggested nuclear power) with the electricity flowing along the shaft. Interestingly, buckytubes have curious—and largely unexplored—electrical properties.

The question of powering the vehicles riding the space elevator is difficult, but there are some small lights at the end of the tunnel. Most of the work is done to get the passenger cab out of earth's gravity well; once it reaches the orbiting station, it could continue outward along the counterweight cable extending into space. If so there is no longer any requirement for power; in fact the cab will now accelerate outward under the influence of the centrifugal force which has now supplanted gravity. Like the skater at the end of a crack-the-whip, the cab will travel faster and faster, until by the time it reaches the outer end of the cable (in its full-length version) it will be travelling eleven kilometres a second. And here's the energy twist: if that car were to be slowed by an electric motor, the electricity generated could be used to power up the next car, and the net cost of energy would be zero.

A unique project like this that bridges earth and space unfortunately brings with it the hazards of both. What happens if your car falls out of the elevator shaft (admittedly an unlikely possibility)? It pays to remember here that while the top of the shaft in geostationary orbit seems motionless, it is actually travelling faster than your starting point on earth (just as the outside horse in the carousel travels faster than the one on the inside), a velocity you begin to acquire as you rise higher. If you fall when you are only a few kilometres above the earth, you will fall down—straight down. If, however, you become detached at an altitude of 25,000 kilometres, you will go into a highly elliptical orbit. Higher still and your orbit will be closer to circular.

If you dare to continue out beyond geostationary orbit, say to 47,000 kilometres, and then lose contact with the cable, you are now heading into space. If you so desire, you have enough velocity to visit the planets, or, as Arthur C. Clarke points out, you have become a planet yourself.

Potentially satellites, asteroids, or, closer to earth, airplanes could collide with the cable. A multistranded cable would likely survive the impact of a small space rock or piece of space junk from a satellite or spent rocket. I mentioned before that severing the cable close to the ground would simply cause the bottom segment to fall to the ground (I guess "simply" might not be the best word to describe the collapse of 35,000 feet of metallic cable, although Jerome Pearson used the words "fail gracefully"). However, making sure commercial airline routes avoided the area of the space elevator would be a pretty elementary first step. There has been some worry that hurricanes would represent a major threat, but hurricanes are rare in the five degrees on either side of the equator where the space elevator would be built.

Jerome Pearson pointed out that tornadoes or strong sustained winds would not be a problem either, because they are not powerful enough to exceed the cable's built-in capacity to withstand tension.

But Pearson did worry about tidal forces exacerbating inherent vibration in the cable and was able to assure himself (if not his skeptics) that the problem was minor. Arthur C. Clarke was concerned about the well-known fact that geostationary satellites wander in orbit because the earth's magnetic field is not uniform, a problem that isn't catastrophic for the satellite but would be a major inconvenience (for people on the ground) if they were attached to the earth. Clarke resolved his own problem by pointing out there are two nodes of stability, one in the Indian Ocean directly above the island of Gan, which was used by the British as a stopover for rockets being transported to Australia in the early years of their space programme. The other stable location is above the Galapagos Islands, which conjures up the horrifying image of Darwin's evolutionary cradle serving as a ground station for space-travelling tourists.

It may be feasible to build space elevators on earth, but it would be much easier elsewhere. Mars is perfect, because it has a gravitational force only about a third of earth's and a handy satellite, Deimos, orbiting just slightly above stationary orbit. Now if you just could tug it down into the right orbit, it would serve as a massive natural counterweight for an elevator cable and could be mined for the material of the cable itself, because it is chock full of carbon. A NASA study has shown that it would be possible to send spacecraft from Mars all the way to the earth for the same price as getting a satellite merely into low Martian orbit. The plan would be to launch the satellite from the Martian surface, then sling it on a 375-kilometre long tether; the tether propels it out thousands of kilometres where it connects with another tether anchored to the moon Phobos; from there to Deimos and thence to earth. (Arthur C. Clarke's 1978 novel, *The Fountains of Paradise,* contains a novel proposal for avoiding the catastrophe of the moon Phobos slicing through a Martian space elevator: the elevator cable would be triggered to vibrate at one of its natural frequencies—like a guitar string—so that it would bend to one side just as the moon passed by.)

A little closer to home the same Jerome Pearson who proposed the orbiting factory for extruding cable sees the moon as an ideal place for space elevators that would reach out to what are called the Lagrange points, places in outer space between earth and moon where the gravitational pulls of the two are balanced, rendering these areas of calm in a sea of gravity. (You can't just put a satellite in orbit around the moon—it will always feel the earth's gravity too.) But a lunarstationary satellite positioned either at the Lagrange point called L4 (running just ahead of the moon in the moon's orbit) or L5 (trailing the moon) would not be susceptible to wandering, as similar satellites orbiting earth can be. Good places for the spas of the 2200s. (My second favourite space idea, the late Gerard O'Neill's giant space colonies, were also to be placed at the Lagrange points.)

As a final note let me bring together two of the great thinkers of this century, Arthur C. Clarke (whose reputation for canny speculation has even survived his forays into television) and Buckminster Fuller.

In the late 1970s Clarke had the bright idea of linking communications satellites together in space as one way of preventing collisions in an ever-more-crowded geostationary orbit. This led to the concept (as he wrote, "Why stop there?") of a continuous habitable structure girding the earth, a "ring city." It would be home to all the earth's communications satellites, deep space missions would be launched from it and it would likely have its own railroad, circling above the earth. Of course you'd get to it by one of many space elevators. When Clarke put this whole vision together he saw the earth as the hub of a gigantic wheel, with people living on the rim or at the hub; the distinction between the two would be blurred. However, Clarke had been anticipated by Buckminster Fuller, who, in 1951, envisioned a "halo-bridge" for the earth, with "Earthian traffic vertically ascending to the bridge, revolving and descending at preferred Earth loci."

Isn't it amazing what such thinkers can do with a concept like space elevators? The elevators are so technically challenging that in all likelihood they will never be built, but Clarke and Fuller, with the impatience of restless minds, are already way beyond them. With thinking like that, who needs Star Trek?

Bibliography

I Just Had to Laugh

Black, Donald W. "Pathological laughter." *Journal of Nervous and Mental Disease* 170 No. 2 (February 1982): 67–71.

Fried, I. et al. "Electric current stimulates laughter." *Nature* 391 (12 February 1998): 650.

Provine, Robert R. "Laughter." *American Scientist* 84 (January/February 1996): 38–45.

Ramachandran, V.S. "The evolutionary biology of self-deception, laughter, dreaming and depression: some clues from anosognosia." *Medical Hypotheses* 47 (1996): 347–362.

Seeing Things

Larson, L.L., trans.. *The King's Mirror*. London: Oxford University Press, 1917.

Lehn, W.H. and I. Schroeder. "The Norse merman as an optical phenomenon." *Nature* 289 (29 January 1981): 362–366.

Bibliography

Sawatzky, H.L. and W.H. Lehn. "The arctic mirage and the early North Atlantic." *Science* 192 (25 June 1976): 1300–1305.

Sane in an Insane World

Rosenhan, D.L. "On being sane in insane places." *Science* 179 (19 January 1973): 250–258.

Various authors' responses to Rosenhan. *Journal of Abnormal Psychology* 84 No. 5 (October 1975): 433–474.

Various authors' responses to Rosenhan. *Science* 180 (27 April 1973): 356–369.

The Barmaid's Brain

Bennett, Henry. "Remembering drink orders: the memory skills of cocktail waitresses." *Human Learning* 2 (1983): 157–169.

Hecht, Heiko and Dennis R. Proffitt. "The price of expertise: effects of experience on the water-level test." *Psychological Science* 6 No. 2 (March 1995): 90–95.

The Invention of Thievery

Fisher, James and R.A. Hinde. "The opening of milk bottles by birds." *British Birds* 42 (1949): 347–357.

Lefebvre, Louis. "The opening of milk bottles by birds: Evidence for accelerating learning rates but against the wave-of-advance model of cultural transmission." *Behavioural Processes* 34 (1995): 43–54.

Sherry, David F., and B.G.Galef, Jr. "Cultural Transmission without Imitation: Milk Bottle Opening by Birds." *Animal Behaviour* 32 No. 3 (1984): 937–938.

Bibliography

The Plant that Rolls

Kirk, David L. and Jeffrey F. Harper. "Genetic, biochemical and molecular approaches to *Volvox* development and evolution." *International Review of Cytology* 99 (1986): 217–289.

Smith, Gilbert M. "A comparative study of the species of *Volvox*." *Transactions of the American Microscopical Society* LXIII No.4 (October 1944): 265–310.

Consumed by Learning

Collins, Harry and Trevor Pinch. *The Golem: What Everyone Should Know about Science.* Cambridge, England: Cambridge University Press, 1993.

McConnell, J.V. "Cannibalism and memory in flatworms." *New Scientist* 21 (February 20, 1964): 465–468.

Rilling, Mark. "The mystery of the vanished citations." *American Psychologist* 51 No. 6 (June 1996): 589–598.

Why Do Moths Fly to Lights?

Callahan, Philip S. "Moth and candle: the candle flame as a sexual mimic of the coded infrared wavelengths from a moth sex scent (pheromone)." *Applied Optics* 16 No.12 (December 1977): 3089–3102.

Fabre, J.H. *The Insect World of J. Henri Fabre.* New York: Dodd, Mead and Company, 1949.

Langmuir, Irving. "The speed of the deer fly." *Science* 87 (1938): 233–34.

McMasters, John H. "The flight of the bumblebee and related myths of entomological engineering." *American Scientist* 77 (March/April 1989): 164–169.

Bibliography

Homo Aquaticus

Hardy, Sir Alister. "Was man more aquatic in the past?" *The New Scientist* (March 17, 1960): 642–645.

Langdon, John H. "Umbrella hypotheses and parsimony in human evolution: a critique of *The Aquatic Ape Hypothesis.*" *Journal of Human Evolution* 33 (1997): 479–494.

Morgan, Elaine. *The Aquatic Ape Hypothesis.* London: Souvenir Press, 1997.

Roede, M., J. Wind, J. Patrick and V. Reynolds, eds. *The Aquatic Ape: Fact or Fiction?* London: Souvenir Press, 1991.

Saint Joan

Butterfield, John and Isobel Butterfield. "Joan of Arc: A medical view." *History Today* 8 (1958): 628–633.

Foote-Smith, Elizabeth and Lydia Bayne. "Joan of Arc." *Epilepsia* 32 No. 6 (1991): 810–815.

Lang, Andrew. *The Maid of France.* New York: Longmans, Green, and Co., 1909.

Henker, Fred O. "Joan of Arc and DSM III." *Southern Medical Journal* 77 No. 12 (December 1984): 1488–1490.

The Effect of Witchcraft on the Brain

Caporael, Linnda R. "Ergotism: The satan loosed in Salem?" *Science* 192 (2 April 1976): 21–26.

Matossian, Mary K. "Ergot and the Salem witchcraft affair." *American Scientist* 70 (July–August 1982): 355–357.

Spanos, Nicholas P. and Jack Gottlieb. "Ergotism and the Salem village witch trials." *Science* 194 (24 December 1976): 1390–1394.

Bibliography

The Vinland Map

Cahill, T.A. et al. "The Vinland Map, revisited: New compositional evidence on its inks and parchment." *Analytical Chemistry* 59 (1987): 829–835.

McCrone, W.C. "The Vinland Map." *Analytical Chemistry* 60 (1988): 1009–1018.

Skelton, R.A., Thomas Marston and George D. Painter. *The Vinland Map and the Tartar Relation*. New Haven and London: Yale University Press, 1995.

The Burning Mirrors of Syracuse

Knowles Middleton, W.E. "Archimedes, Kircher, Buffon and the burning mirrors." *Isis* 52 (1961): 533–543.

Mills A.A., and R. Clift. "Reflections on the 'burning mirrors of Archimedes.'" *European Journal of Physics* 13 (1992): 268–279.

Simms, D.L. "Archimedes and the burning mirrors of Syracuse." *Technol. Culture* 18 (1977): 1–24.

The Monks Who Saw the Moon Split Open

Calame, Odile and J. Derral Mulholland. "Lunar crater Giordano Bruno: A.D. 1178 impact observations consistent with lunar ranging results." *Science* 199 (24 February 1978): 875–877.

Hartung, Jack B. "Was the formation of a 20-km-diameter impact crater on the Moon observed on June 18, 1178?" *Meteoritics* 11 (1976): 187–194.

Nininger H.H. and G.I. Huss. "Was the formation of lunar crater Giordano Bruno witnessed in 1178? Look again." *Meteoritics* 12 (1977): 21–25.

Bibliography

Antlion King

Eisner, Tom, Ian Baldwin and Jeffrey Conner. "Circumvention of prey defense by a predator: Ant lion vs. ant." *Proceedings of the National Academy of Sciences USA* 90 (July 1993): 6716–6720.

Lucas, Jeffrey R. "The biophysics of pit construction by antlion larvae (*Myrmeleon*, Neuroptera)." *Animal Behaviour* 30 (1982): 651–664.

Topoff, Howard. "The pit and the antlion." *Natural History* (April 1977): 65–71.

Wheeler, William Morton. *Demons of the Dust.* New York: W.W. Norton and Company Inc., 1930.

The Bacteria Eaters

Karam, J.D. ed. *Molecular Biology of Bacteriophage T4.* Washington DC: American Society for Microbiology, 1994.

An Uneasy Bargain

Gabriel, S. et al. "Cystic fibrosis heterozygote resistance to cholera toxin in the cystic fibrosis mouse model." *Science* 266 (7 October 1994): 107–109.

Pier, G. et al. "Salmonella typhi uses CFTR to enter intestinal epithelial cells." *Nature* 393 (7 May 1998): 79–82.

Schroeder, S.A. "Protection against bronchial asthma by CFTR ΔF508 mutation: A heterozygote advantage in cystic fibrosis." *Nature Medicine* 1 No. 7 (July 1995): 703–705.

A Silver Lining

Griffith Smith, Neal. "The advantage of being parasitized." *Nature* 219 (17 August 1968): 690–694.

Soler, M. et al. "Magpie host manipulation by great spotted cuckoos: evidence for an avian Mafia?" *Evolution* 49 No. 4 (1995): 770–775.

Bibliography

Tee Time at the Royal Institution

Lord Rayleigh. *The Life of Sir J.J. Thomson*. London: Cambridge University Press, 1942.

Thomson, Sir J.J. and Sir James Crichton-Brown. "The Dynamics of a Golf Ball." In *The Royal Institution Library of Science, Physical Sciences*. Vol. 7, edited by Sir William Lawrence Bragg and George Porter. Barking, Essex: Elsevier Publishing Company, 1970.

It'll Practically Go Forever

Angrist, Stanley W. "Perpetual motion machines." *Scientific American* (January 1968): 114–122.

Ord-Hume, Arthur W.J.G. *Perpetual Motion: The History of an Obsession*. London: Allen & Unwin, 1976.

Yam, Philip. "Exploiting zero-Point energy." *Scientific American* (December 1997): 82–85.

Going Up?

Clarke, Arthur C. "The space elevator: 'Thought experiment', or key to the universe?" Address to the xxxth International Astronautical Congress in Munich, 20 September 1979. In *Advances in Earth Oriented Applied Space Technologies*. London: Pergamon Press Ltd. 1981. Also at http://plains.uwyo.edu/~mickray/clark.htm

Isaacs, J.D. et al. "Satellite elongation into a true "sky-hook." *Science* 151 (1966): 682–683.

Forward, Robert. "Magic beanstalks." In *Future Magic*, 53–78. New York: Avon Books, 1988.